Exposing the Myths of Industrial Precision Measurement Control

By
Richard Clark

Hanser Gardner Publications
Cincinnati, OH

Library of Congress Cataloging-in-Publication Data

Clark, Richard, 1966-
  Exposing the myths of industrial precision measurement control / by Richard Clark.
    p. cm.
  Includes bibliographical references and index.
  ISBN 1-56990-393-X
  1. Engineering instruments--Calibration. 2. Mensuration. I. Title.
  TA165.C55 2006
  620'.0044--dc22
                    2006007669

While the advice and information in *Exposing the Myths of Industrial Precision Measurement Control* are believed to be true, accurate, and reliable, neither the author nor the publisher accept any legal responsibility for any errors, omissions, or damages that may arise out of the use of this advice and information. The author and publisher make no warranty of any kind, expressed or implied, with regard to the material contained in this work.

A ***Modern Machine Shop*** book published by
Gardner Publications, Metalworking's Premier Publisher
www.mmsonline.com

Hanser Gardner Publications
6915 Valley Avenue
Cincinnati, OH 45244-3029
www.hansergardner.com

Copyright © 2006 by Hanser Gardner Publications. All rights reserved. No part of this book, or parts thereof, may be reproduced, stored in a retrieval system, or transmitted in any form or by any means without the express written consent of the publisher.

Introduction

"But when a long train of abuses and usurpations, pursuing invariably the same Object evinces a design to reduce them under absolute Despotism, it is their right, it is their duty, to throw off such Government, and to provide new Guards for their future security." —Taken from the Declaration of Independence, 4 July 1776

One day a man noticed a crack on the wall of a bedroom in his relatively new home. Wanting to keep the home looking new, the man called a painter to have the crack repaired. The painter filled the crack with putty, sanded the surface, and repainted the entire wall so no one would notice the repair had been made. The painter submitted the bill and the man paid him. The painter was happy, the man was happy, all was well.

About a month later, the crack reappeared on the wall. The man contacted the painter and expressed his disappointment; the painter came over to the house and repaired the crack again at no charge. The man was happy, and all was well.

About 2 weeks later, the crack reappeared on the wall with several other cracks around it. The man decided to find another painter, whom he called over for an estimate. The painter examined the wall, looked at the man and said… "I'm sorry, I can't help you." The man said "wait a minute, you're a painter, right, and I've got this problem with cracks on my wall. Why can't you help me?" The painter said "you don't have a problem with cracks on your walls." He explained to the man there were indeed cracks on the wall, but that was not his problem. The man's problem was the foundation of the house was shifting and the painter would be robbing that man of his money to putty and paint the cracks because as soon as the foundation shifted again, the cracks would reappear. He said to the man "… you see, until you fix your foundation you will always be fixing cracks on your walls."

Dr. Tony Evans – 1995
President of The Urban Alternative
www.tonyevans.org

In today's industrial world of standards I see many facilities that try and fix shifting foundations by repairing cracks on the walls. The goal of every industrial facility is to exceed their customers' expectations for quality in the products they produce. The walls surrounding these products are the quality system of the facility that makes them, and the foundation of that system is the measurements (or control of the measurements) used to determine the status of the quality.

Too often, a facility wants to receive an ISO, QS, or TS certification, yet they want to keep "flexibility" in their systems to allow them to keep doing things the way they did before the certification. This equates with wanting to steal second, while keeping your left foot on first. The end result sometimes becomes a tangled web of procedures that we say we do (but really don't), that do not produce the required results. It is penny wise and dollar foolish.

This book will take you along the path I was fortunate to travel over the last nine years. What is taught in this book will often seem to reside far outside the box. This is by design. We'll use the cracks on the wall and shifting foundation philosophy as we learn what national and international standards say, and don't say, about control of inspection, measurement, and test equipment. We'll discuss how to design a strong system that meets both the letter (and the intent) of these standards. We'll also discuss the common suggestions made from the "can't we just?" library and why they don't work. Most importantly, we'll discuss how to design, develop, and implement an inspection, measurement, and test equipment control system from the ground up without doing anything more difficult than counting to 10 with beads on a string. You see, if it were much more difficult than this, yours truly couldn't teach it.

Richard Clark – 2006

Contents

Introduction iii

Acknowledgments and References vi

Chapter 1:   Why Control Instruments? 1

Chapter 2:   Artifact Standards 9

Chapter 3:   Protocol Standards 17

Chapter 4:   Getting Started 27

Chapter 5:   Common Outsourced Calibrations 37

Chapter 6:   Scheduling Outsourced Calibrations:
             Tips and Tricks that Can Save Time and Money 47

Chapter 7:   Learning, Understanding, and Conquering Thermal Effects
             in Industrial Measurement 53

Chapter 8:   What You Can't (Not) Know about SPC 63

Chapter 9:   Introduction to Measurement Systems Analysis
             (Gage Repeatability and Reproducibility) 73

Chapter 10: Elements of a Measurement 87

Chapter 11: "Know the Uncertainty" is Not an Oxymoron
             (an Introduction to Uncertainty Concepts and Contributors) 95

Chapter 12: Uncertainty Part Two: Budgets and Estimation Concepts 103

Chapter 13: Selecting and Implementing Software 113

Chapter 14: Developing a Smooth Flow 123

Chapter 15: Out-of-Calibration Situations 135

Chapter 16: The Gage is Calibrated and Issued,
             But the Readings Don't Seem to be Right 141

Chapter 17: Evaluating Training Needs and Concepts 149

Chapter 18: Training "By the Numbers" 159

Chapter 19: Putting it All Together 171

Index 177

Acknowledgements

The author would like to thank the following organizations that contributed both directly and indirectly to the completion of this book.

Business Process Improvement
www.bpiconsulting.com/
www.spcforexcel.com/

Cincinnati Precision Instruments
www.cpi1stop.com

CMM Technology, Inc.
www.cmmtechnology.com

The Dyer Gage Company
www.dyergage.com

Ertco Precision

Glastonbury Southern Gage
www.gsgage.com

HN Metrology Consulting
www.hn-metrology.com/

Leimer Engineering

Mitutoyo America
www.mitutoyo.com/

Math Options Inc.
Designers and distributors of Easy Gage R&R software
mathoptions.com/

NWCI Calibration and Inspection

Prompt Consultants Inc.
www.promptconsultants.com/

Performance Tool Inc.

Tru-Stone Technologies

The author would like to thank the following individuals whose support over the years helped guide him along the path that led to the completion of this book.

Ramona Clark - Portland, IN

Betty Jean Means - Columbus, OH

SFC (Retired) Thomas J. Ravenell - Raleigh, NC

LTC Walt Pjetraj – Columbia, SC

Jim Elbert – Atlanta, GA

Camille Elick, FNP

Gail Franks

Duane Weaver

Warren Lowe - Portland, IN

Sara Farris – Portland, IN

Denise Craybill

Gary Smith - Portland, IN

Crystal Laux - Portland, IN

Harold Smith - Dunkirk, IN

Ed Bennet - Dunkirk, IN

Claudia Sneed - Portland, IN

Scott C. Beavers - Cincinnati, OH

Pat Lawrence - San Clemente, CA

Garth Taggart - San Clemente, CA

James Vannoy - San Clemente, CA

Steve Plymire – Cincinnati, OH

Mike Dukehart

Jerry Guffy

Matt Dye

Dave Schwab

Mr. Suga

Gordon (Gordie) Skattum

Terry Davis

Bonnie Maitlen

Rob and Susan Moser (for their kind support and contributions to Chapter 19)

Joe McKenna - Cleveland, OH

Pete Zelinski - Cincinnati, OH

Jim Lorinz - Solon, OH

Woody Chapman

Glynn and Julie Barber – Portland, IN

References

The author would like to mention the following material that contributed both directly and indirectly to the completion of this book or the inspiration to write it.

Winning the Race: The Greg Moser Story - Burns, C. DDS Publishing (1998)

Cliff's Quick Review: Algebra - Wiley Publishing Inc.

Cliff's Quick Review: Geometry - Wiley Publishing Inc.

Cliff's Quick Review: Physics - Wiley Publishing Inc.

Cliff's Quick Review: Statistics - Wiley Publishing Inc.

The Romance of Metrology - An excerpt from Calibration: Philosophy in Practice by the Fluke Corporation

QS-9000: 1998 – Automotive Industry Action Group (A.I.A.G.)

Measurement System Analysis: 3rd Edition – Automotive Industry Action Group (A.I.A.G.)

Statistic Process Control - Automotive Industry Action Group (A.I.A.G.)

Evaluating the Measurement Process – SPC Press, Inc.

Statistical Methods for the Process Industries - Quality Press, 1991

SPC for Microsoft Excel Reference Manual - Business Process Improvement

ISO/IEC 17025 General requirements for the competence of testing and calibration laboratories

ISO 10012-1 Quality assurance requirements for measuring equipment - Part 1: Metrological confirmation system for measurement equipment

Chapter 1
Why Control Instruments?

*"Inspection measuring and test equipment shall be used in a manner that ensures the **measurement uncertainty is known** and consistent with the required measurement capability."*—QS-9000:1998

Control, when it really comes down to it, is somewhat an illusion. We never have complete control of anything; there are always unknowns within the constants. If you think about it, control comes down to using constants to keep up with the changes (or change possibilities) of the unknowns. Always remember that for every cause there is an effect, and for every action, a reaction.

To develop a clear understanding of Metrology (measurement science), we must first look at internationally accepted quality standards and interpret them in order to clarify their meaning. We will use the QS-9000:1998 standard which may seem obsolete in some circles—due to the more current TS-16949 standard—but the QS-9000 standard is (in this author's opinion) the best standard by far to use as a model when developing an industrial in-house measurement control system. Whenever the words "the supplier" appear in the QS-9000 standard, we can replace them with "your facility." In the tutorial that follows, we will discuss the QS-9000 standard, which is a protocol or "paper" standard that is a published guideline we must comply with. In the language of metrology and calibration, we often refer to a standard as an artifact of documented known value.

To begin, it will be helpful to clarify the often misunderstood protocol standard QS-9000:1998 Element 4.11, which states:

The supplier shall establish and maintain documented procedures to control, calibrate and maintain inspection, measuring, and test equipment used to demonstrate the conformance of product to the specified requirements.

This means that when a customer requires a part to be made according to an engineering drawing, all specifications listed on the drawing become "specified requirements" and **must** be measured and confirmed. All instruments used to confirm these requirements must be "controlled instruments," meaning a facility must have and maintain documented procedures to ensure that the instruments are kept under control. In other words, the instruments must be calibrated and maintained, and the facility must possess procedures that document the entire process as can be seen in this passage from the standard.

The supplier must identify all inspection measuring and test equipment that can effect product quality, and calibrate and adjust them at prescribed intervals, or prior to use, against certified equipment having a known valid relationship to international or national recognized standards.

This means the facility must have a system to identify all of the equipment (controlled instruments) that will be or can be used to confirm the features (specified requirements), as well as any other instruments that can affect the quality of the product. This can be explained with a simple example: If a customers' documentation (part drawing, control plan, etc.) requires dimension #1 to be confirmed using a micrometer, your system must be able to identify this micrometer as a controlled measuring instrument. Your facility may decide (for internal reasons) to confirm the dimension with additional instruments. Even though the additional instruments are not a customer requirement, these instruments could affect product quality if an operator were to measure the dimension (using the additional instruments) and make an off-set to the machine that produces the part.

This equipment must be calibrated and adjusted at defined intervals or prior to use. The calibration must compare the instruments (being calibrated) to certified equipment or artifacts that are traceable to national or international accuracy standards. This process is called **traceability**, and enforcement is very simple.

If, for example, I have a wrist watch and want to see how accurate it is, I might compare it to the clock displayed on CNN. If my watch is within a second of CNN's clock, that's great, but how can I be 100% sure CNN's clock is accurate? CNN's clock is, most likely, compared to a clock which is compared to another clock, which is compared to the atomic clock at N.I.S.T. (the acronym for the National Institute of Standards and Technology, formerly NBS–National Bureau of Standards) this official U.S. time can be seen at www.time.gov.

Another great example is the digital scale found at every deli in every supermarket in America. Somewhere on the scale is an inspection label from the State Bureau of Weights and Measures. The scale was calibrated using certified weights or masses. The masses used were compared to other masses (of greater accuracy) that were (by traceability) compared to masses controlled by N.I.S.T. Think of traceability as a chain that, no matter how long, must remain unbroken.

The standard further states:

Determine the measurements to be made and the accuracy required and select the appropriate inspection, measuring and test equipment that is capable of the necessary accuracy and precision.

The measurements that are performed within a quality system must be documented and predetermined. A facility must also decide and document how accurate each measurement result must be to be considered acceptable. These measurements must be taken using measurement and test equipment that is capable of measuring to the documented accuracy and precision required by our acceptance criteria.

It may seem like we just talked ourselves in a circle, but this is not the case. There are some fundamental, you might even call them sacred, rules of metrology that, if followed, will almost never let you down. One of these is called the "rule of ten" or the 10:1 ratio. This is a guideline that should be followed whenever possible and practical. We'll discuss this concept and when it is and is not practical in greater detail later in this book, but it's worth touching on now.

When an accuracy level (tolerance) is determined for an application, the measurement instrument or system should be accurate and precise within a ratio 10:1 of the tolerance for the application. It's best described through examples. Examples in this book will use Metric and/or inch explanations.

- Inch example: A length measurement specified on a customer's drawing (or print) is listed as 2.000" ± .005.

- Metric example: A length measurement specified on a customer's drawing (or print) is listed as 50mm ± 0.15.

If we decided to confirm this characteristic using a digital or digimatic caliper, we would be using an instrument that, according to its manufacturer, is accurate within ±.001" (± 0.0254mm). This information is seen in *Figure 1-1*. What this means in layman's terms is that if two "identical" brand new calipers were issued to separate locations on the shop floor, caliper A could consistently measure +.001" (+0.025mm) and caliper B could consistently measure -.001" (-0.025mm) on the same part. The calipers could differ -.002" (0.05mm) from each other, and still be accurate within accepted accuracy standards. Using the example listed above we can determine if the inaccuracy of our instrument fits within a 10:1 ratio (or below 10%) of the length specification.

- Inch example: A length measurement specified on a customer's drawing (or print) is listed as 2.000" ± .003.

Total tolerance = .006"

Possible instrument error at calibration = .001"

Possible instrument error consumes 16.7% of the part tolerance.

- Metric example: A length measurement specified on a customer's drawing (or print) is listed as 50mm ± 0.1.

Total tolerance = 0.2mm

Possible instrument error = 0.03mm

Possible instrument error consumes 15% of the part tolerance.

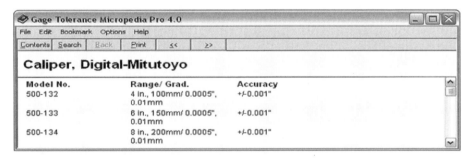

Figure 1-1: The accuracy of the caliper is ± .001" (± 0.0254 mm). Image courtesy Gage Tolerance Micropedia by Prompt Consultants, Inc.

However, if we had chosen a digital micrometer for our measurement as shown in *Figure 1-2*, we would be using an instrument that, according to the manufacturer, was accurate within ±.0001" (± 0.00254mm). This information is seen in *Figure 1-3*. If we plug in the possible error of this instrument, our scenarios would look much better.

- Inch example: A length measurement specified on a customer's drawing (or print) is listed as 2.000" ± .003.

Total tolerance = .006"

Possible instrument error = .0001"

Possible instrument error consumes 1.7% of the part tolerance.

- Metric example: A length measurement specified on a customer's drawing (or print) is listed as 50mm ± 0.1.

Total tolerance = 0.2mm

Possible instrument error = 0.002mm

Possible instrument error consumes 1.0% of the part tolerance.

One instrument type (digital micrometer) complies with the 10:1 rule and one instrument type (caliper) does not. This topic will be applied in more detail later in this book.

Figure 1-2: A digital micrometer. Image courtesy of Mitutoyo America.

Micrometer, Outside-Mitutoyo

Model No.	Range/ Grad.	Accuracy
101-103	0-25mm/ 0.01mm	+/-0.002mm
101-104	25-50mm/ 0.01mm	+/-0.002mm
101-105	0-1"/ 0.001"	+/-0.0001"
101-106	1-2"/ 0.001"	+/-0.0001"
101-107	0-25mm/ 0.01mm	+/-0.002mm
101-109	0-1"/ 0.001"	+/-0.0001"
101-110	1-2"/ 0.001"	+/-0.0001"

Figure 1-3: The accuracy of this micrometer is ± 0.002 mm. Image courtesy Gage Tolerance Micropedia by Prompt Consultants, Inc.

Now we will have a look at other quotes from the standard (each in italics below) and discuss their implications.

Define the process employed for calibration of measuring equipment, including the details of equipment type, unique identification, location, frequency of checks, acceptance criteria and the action to be taken when the results are unsatisfactory.

To comply with this requirement, a system must define the procedures and methods used when calibrating each type of instrument. These procedures should specify the measurements (or series of measurements) that will be taken on each type of equipment. These procedures must include: instrument identification, location of equipment, how often it is calibrated, how accurate each instrument must be to "pass" the calibration, and what actions will be taken when instruments do not "pass" a calibration.

Identify inspection measuring and test equipment with a suitable or approved identification record to show the calibration status.

All inspection, measuring, and test equipment that is calibrated, controlled, and approved to confirm product quality must be identified to clearly state the status of its most previous calibration. The most practical and effective method to do this is to use calibration labels as seen in *Figure 1-4*. Some facilities use the unique instrument ID number (serial number) along with an instrument master list to identify the calibration status of equipment. Other facilities may use color coding along with a documented schedule of when instruments with certain colors are due for calibra-

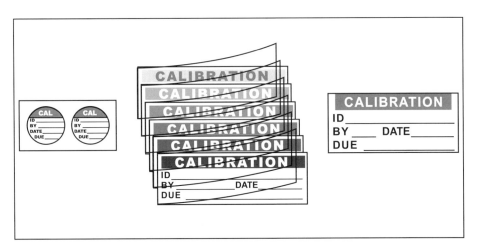

Figure 1-4: Calibration labels. Image courtesy of Advanced Calibration Label.

tion. All of these methods present a facility with different constants and variables. You may want to use the trial and error method to decide which method is best for your facility.

Maintain calibration records for inspection, measuring and test equipment.

On the surface, this sounds self-explanatory. However, calibration records should include as much information as possible from the test. The protocol standards and artifact standards used during calibration, data of the values obtained when the instrument being calibrated was used to measure the standards, environmental data, and any other pertinent information.

Access and document the validity of previous inspection and test results when inspection, measuring or test equipment is found to be out of calibration.

This requirement has the potential for devastation. Most people think it simply means that if an instrument is found to be outside of accuracy requirements during calibration, the previous calibration result must be confirmed. Not quite. It specifies the "validity of previous inspection and test results" as opposed to only the validity of previous calibration(s). The accuracy of the instrument during a previous calibration is where the suspect range begins. The investigation must start there and move forward in time to validate the inspection results of **all** the products the instrument in question confirmed since the last known (in tolerance) calibration. Appropriate action must be taken should the quality of any products be in question. This *could* (in a worse case) ripple out to a product recall.

Ensure that the handling, preservation and storage of inspection, measuring and test equipment is such that the accuracy and fitness for use is maintained.

The calibration process documents that an instrument was tested within a stated specification, on a given date, in a controlled environment—period! It is never a guarantee the instrument will "stay good" until the next scheduled calibration due date. Common sense in the handling, preservation, and storage of equipment will help to maintain the accuracy present during calibration. Most inspection, measurement, and test equipment has accompanying documentation that specifies proper storage, use, and handling to maintain the accuracy and fitness for use.

I often compare this section of the QS-9000 standard to a physical wellness plan that most health agencies recommend we all follow. I can visit my doctor and undergo an in-depth physical or wellness check that documents my state of health (or wellness) at the time of the test, but it's up to me to make wise lifestyle choices in my diet, exercise, etc. to maintain this level of wellness. Some facilities think just because an instrument has a label stating it was calibrated on a certain date and not due for another check until a future certain date, it is guaranteed good until this due date. This would be the equivalent of eating nothing but doughnuts, drinking nothing but soft drinks, and refusing to exercise because my wellness results said "I'm good and not due for a recheck for another year."

Summary: The intent of the QS standard is to require a facility to implement and maintain a system in which inspection equipment is calibrated (and maintained) to a specified accuracy using equipment that is certified to an accepted national or international standard. The calibration results are documented and the instruments are identified as calibrated for use. The controlled instruments are used to confirm the features (specified requirements) as identified by a customer approved drawing or other similar document.

Chapter 2
Artifact Standards

"Fast is fine, but accuracy is everything!"—Wyatt Earp

Before any internal standards can be developed or even considered, the definition(s) of the word "standard" must be clearly understood. When dealing with systems that control inspection, measurement, and test equipment, the word standard can have one of two different meanings depending on the context in which it is referred.

The first type of standard we will discuss is the **artifact standard.** This is a specific object or phenomenon that represents and defines a known or accepted value. In an industrial facility this is found in the form of the "grandfather" gage block set that is sent out for calibration to an outside inspection laboratory. These gage blocks are then used to compare values obtained by instruments and objects of unknown value (or accuracy) in order to document, correlate, adjust, or account for the deviation between the instruments of unknown value (or accuracy) and the objects of known value.

When artifact standards are sent to an outside laboratory they are compared to artifacts of a greater accuracy, such as an outside laboratory's gage blocks. The accuracy of the outside laboratory's gage blocks is determined when they are sent to another facility for comparison to artifacts of an established accuracy. This documented process is called traceability. In the United States, most if not all measurements are eventually traced to the National Institute of Standards and Technology (N.I.S.T.).

When you think about it, traceability to a standard (and eventually to N.I.S.T.) is all around us. When you go to the gas pump at your local convenience store you'll see a label attached to the pump documenting that the accuracy of the pump was certified by the State Bureau of Weight and Measures. When the digital display states the pump has given you 1.000 gallon of gasoline, it has been confirmed to be within a certain accuracy tolerance of 1.000…0 gallons. The standard used was a certified container from the State Bureau of Weights and Measures that is periodically compared to containers of greater accuracy that eventually trace to N.I.S.T.

If you need a demonstration model of the purest from of traceability, you need not look any further than any physics textbook or website. The International System of Units (SI) consists of seven base units:

1. Length: The Meter, defined as the length of the path traveled by light in a vacuum during a time interval of 1/299,792,458 of a second.

2. Time: The Second, defined as the duration of 9,192,631,770 cycles of radiation corresponding to the transition between two hyperfine levels of ground state of the cesium~133 atom.

3. Electric Current: The Ampere, defined as the electric current producing a force of 2×10^{-7} newtons per meter of length between two long wires, one meter apart in free space.

4. Thermodynamic Temperature: The Kelvin, defined as 1/273.16 of the thermodynamic triple point of water.

5. Luminous Intensity: The Candela, defined as the luminous intensity in a given direction of a source that emits monochromatic radiation at a frequency of 540×10^{12} hertz with a radiant intensity in that direction of 1/683 watts per steradian.

6. Amount of Substance: The Mole, defined as the amount of substance of a system that contains as many elementary items as there are atoms in 0.012 kilogram of carbon 12.

7. Mass: The Kilogram, the only unit defined as a physical artifact, a mass equal to the mass of the international prototype of the kilogram, which is an artifact cylinder of platinum iridium alloy kept by the BIPM near Paris, France.

I would not claim to be an atomic scientist, but one thing I can determine about traceability by reviewing the definitions of the SI

Figure 2-1: The World's standard for mass. Image courtesy of the Bureau International des Poids et Mesures (BIPM) - Paris, France.

units is this: six out of the seven SI units are defined as scientific phenomenon, and one is an actual artifact, which has to mean that every controlled measurement taken by mankind traces back to the 1-kilogram artifact cylinder that would fit in the palm of your (gloved) hand. This covers a measurement using a caliper on the shop floor or an engineer uploading axial vector movements into the navigational unit of the Hubble Space Telescope, and every controlled measurement in-between.

This concept of traceability can be used to develop some very inexpensive standards that can be created and used internally with some in-house measurements and calibrations that, if performed by outside calibration laboratories, would be very expensive. Let me explain. We'll visualize three production lines within a shop. Each production line has a parts washing process for contaminates to be removed before a later assembly process. Each parts washer has a submersed thermometer in the washing and rinsing areas of the machine, and these temperatures are referenced on the customer approved control plan so they must be controlled (calibrated and traceable). We'll use these requirements as examples:

- Wash temperature listed on control plan: 170°F to 190°F
- Rinse temperature listed on control plan: 120°F to 150°F.

These thermometers must be calibrated. We can have an accredited outside calibration facility calibrate these thermometers at a cost of about $100 each. This option will cost $600 for all 3 washers and we will experience machine downtime while the

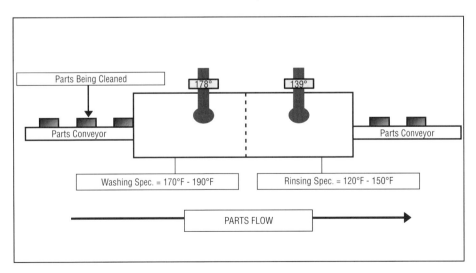

Figure 2-2: A parts washing process for contaminates.

calibration is performed. Another option is to calibrate the thermometers ourselves. How can we do this? We can develop an internal standard using two accepted natural phenomena of science.

When we are in elementary school we learn that water freezes at 32°F (or 0°C), and water boils at 212°F (or 100°C). I once asked the question, if water freezes at 0°C, then at what temperature does ice melt? Is it 0.1°C, 0.01°C, or 0.001°C. It's none of the above. 0°C and 100°C are transfer points when water changes from one form of matter into another. If a glass of water is placed into a freezer, it transfers to ice when it reaches 0°C. And if an ice cube is set on the countertop, the surface of the ice transfers to water when it reaches 0°C. The same concept applies at the boiling point. When water reaches a temperature of 100°C it transfers into steam, and the condensation turns it back into liquid when the steam temperature cools to 100°C. Now we have two known values that we can create in any kitchen.

If we place well-crushed ice into a strainer, the temperature of the water that appears on the surface of the ice as it begins to melt will be 0°C. We can place the probe of the thermometer into the "snow cone" and compare the measured value (what the thermometer reads) to the known value. Because the ice is placed in a strainer, the excess water will drip away leaving only the thawing ice. After this test is performed we can place a pot of water on a stove burner until it reaches a rolling boil. We can then immerse the probe into the boiling water which has a know value of 100°C (I was recently in the pots and pans aisle of a department store and noticed the freezer thermometers had an "ice check" point at 32° and the oven thermometers had a "boil check" at 212°F).

Is this test perfect? No, certainly not; we will most likely see some error (or uncertainty) of 1°C or so in either or both measurements, but we can factor this error (or uncertainty) into our internal protocol standards (discussed in Chapter 3) and use adjusted specifications of:

- Wash temperature listed on control plan: 172°F – 188°F
- Rinse temperature listed on control plan: 122°F – 148°F.

If anyone argues and states this method is not acceptable, then we better call every standards organization in the world and inform them they should reject the SI definition of the meter (the length of the path traveled by light in a vacuum during a time interval of 1/299,792,458 of a second) and the Kelvin (1/273.16 of the thermodynamic triple point of water) as well.

I'm a firm believer in covering all the bases, and sometimes it's easier to take a few extra steps to make sure a standard is met. For this peace of mind, we could have one of the thermometers calibrated by an accredited laboratory and use this as a transfer standard to "confirm" our ice point and boiling point. Then, when we place our thermometer of unknown accuracy into our ice point and boiling point, we could place the calibrated thermometer in right beside it. This should satisfy even the most narrow-minded critic.

Processes that use a measurement called "specific gravity" can demonstrate another example of developing a standard. Ertco Precision (The Ever Ready Thermometer Company) defines a hydrometer (*Figure 2-3*) as an instrument that measures the density of liquids. Although hydrometers actually measure density, measurements are commonly expressed in terms of specific gravity; which is the ratio of the density of the substance being measured to the density of water at the same temperature. In precise measurements, the reference may be to pure (double-distilled) water at 4°C (39.2°F). In engineering practice, the reference frequently is to pure water at 15.6°C (60°F); and the value of unity established for water is 1.000. Thus, liquids with a specific gravity less than 1.000 are lighter than water, and those greater than 1.000 are heavier than water. The specific gravity for a given gas is the ratio of the density of the gas to the density of air. Since the density of air varies markedly with both temperature and pressure, precise measurements should reflect both conditions. Common reference conditions are 0°C at 1 atmosphere.

Hydrometers work on Archimedes principal, which is that the up-thrust on the body immersed in a fluid (either liquid or gas) equals the weight of the fluid displaced. According to this principal, the hydrometer will sink until it displaces its own mass. The volume of liquid displaced is indicated by the level of the scale, and the density is equal to the mass divided by the volume. Since fluid density fluctuates with temperature, proper temperature corrections must be made. A separate thermometer or a combined hydrometer containing its own thermometer should be used to determine the temperature of the sample.

As the Hydrometer floats (or sinks) in the fluid, the stem scale is used to determine how buoyant it becomes in the fluid. If a fluid has a specific gravity of 0.90, we can determine that 1cc of the fluid would have a mass of 0.90 grams. It's approximately 10% "lighter" than water. These hydrometers are usually made of glass and are used to measure fluids. What happens when wet glass objects are handled frequently in production driven environments? I'll give you a hint: Where's the broom and dustpan?

I once worked with a facility that used hydrometers at 10 separate processes. The specific gravity was listed on the customer approved control plan, which meant the hydrometers needed to be calibrated and controlled. The cost of new hydrometers was about $25 each. The cost of accredited calibration and certification was over $100 extra per hydrometer. I suggested we first develop an internal standard (or known value). To do this, we purchased one hydrometer of each range used on the shop floor and paid for each one to be calibrated and certified before delivery. These hydrometers were assigned to the calibration department as master hydrometers (or transfer standards). We then ordered mass quantities of non-calibrated hydrometers for $25 each. Finally, we created sample mixtures of adhesive and thinner to the upper and lower limits of each type (scale range of the hydrometer) and confirmed the mixtures with our calibrated hydrometer. Now we had established known values, which, through the calibration of the master hydrometers, was traceable to national standards.

Another type of artifact standard that can be created in-house is termed a **reference standard**. This can be any object that an instrument is used to measure. I usually recommend using a sample part. A reference standard has an established value that is determined by several sample measurements taken over a specified time frame. These measurements can be taken with a shop floor issued instrument and then, if desired, confirmed through layout inspection type measurements taken in a controlled lab environment with dedicated equipment. How this process works is very simple and will be discussed graphically later in this book. We'll use a concise example to cover it now.

Let's assume for a second that we use micrometers at several similar shop floor processes to confirm and control outside diameter measurements on several customer part types. The micrometers are calibrated within federal accuracy standards (explained in the next chapter) at 6 month intervals. We decide it would be very beneficial to have standards issued to our shop floor processes so the operators can confirm that the accuracy

Figure 2-3: Hydrometer used to measure specific gravity. Image courtesy Ertco Precision.

of their micrometers has not shifted since yesterday, the day prior, or any day since the most recent calibration was performed. If we obtained a sample and marked a fixed diameter across the part, we could measure the diameter three times a day for 25-days using a specific micrometer. Then, after the 25-days we would be able to calculate the 75-reading average. We could compare this average to a measurement taken in our inspection lab using a very accurate bench-top type micrometer. If the variation between the values (our 75-measurement average and bench micrometer reading) was low enough, Presto! We have a reference standard. If an operator suspected his/her micrometer was giving questionable measurement readings, they could measure the reference standard to gage (pun intended) how the micrometer was measuring against our established value (75-reading average).

Understanding how grandfather standards, transfer standards, and established reference standards are used and how each correlates with the others will serve as the framework for your facility's measurement control.

Chapter 3
Protocol Standards

In the beginning, was the Word....

The next category of a standard we will discuss is the protocol or "paper" standard. This type of standard is a document or other record that states a given requirement or series of requirements. These standards come in the form of the QS-9000 and TS-16949 quality standards that industrial facilities are certified to. Another example is the ISO-17025 standard that third party inspection and calibration labs are accredited to. In either of the above scenarios, the facilities involved must have documented internal standards (procedures, work instructions, etc.) that state how certain actions must be performed in order to comply with the standard the facility is accredited to.

When developing and writing internal protocol standards, I would recommend following a safe philosophy of three steps:

1. Say what you do,

2. Do what you say,

3. Prove it!

Your internal standards are a hierarchy. At the top are controlled **procedures**, which are usually about 90% general and 10% specific. Procedures usually describe the "what," but might not go into much detail about "who" or "when." Procedures can simply follow the format (element and paragraph) of the Quality Standard (QS-9000) and restate your facility's policy concerning meeting the letter of the standard.

For example, the QS-9000:1998 standard states:

The supplier shall establish and maintain documented procedures to control, calibrate and maintain inspection, measuring, and test equipment used to demonstrate the conformance of product to the specified requirements.

A procedure might state:

Inspection, measurement, and test equipment is identified, calibrated (and adjusted if necessary) at prescribed intervals as defined by internal work instructions.

Inspection, measurement, and test equipment shall be calibrated and used in an environment controlled in a manner that ensures the integrity of the accuracy is maintained.

Work instructions (however titled) are the next level of controlled standards. These standards tend to be about 90% specific and only 10% general, and explain in detail when certain tasks are performed, who performs them, and who confirms them. Work instructions also describe internal accuracy targets, limits, and acceptance criteria, and how results are to be documented. Both procedures and work instructions must be controlled documents through your facility's document and data control system. Explaining document and data control (QS-9000 element 4.5) would take its own book, so just remember the two items (minimum) **controlled documents should have:**

1. A unique identification or document control number,
2. A revision date.

This requirement does not necessarily mean that procedures and work instructions must be written internally from scratch. OEM documentation can be assigned control numbers and revision dates. If you have access to prewritten work instructions through software or other mediums, a copy can be printed for your records, assigned a control number along with a revision date, and used as your internal standard.

The following is an example courtesy of GMS (Gage Management Systems) for Windows and Gage Tolerance Micropedia from Prompt Consultants, Inc. It is a sample work instruction for the calibration of caliper instruments.

Calipers:

Purpose:

To establish a standard procedure for the calibration of calipers

Scope:

All calipers used to measure, gage, test, inspect and control part compliance to specifications and drawings.

Referenced Documents:

GGG-G-111c Calipers and Gages, Vernier

Materials Required:

Cleaning solution

Hard Arkansas stone

Lint free cloth

Gage oil

Long gage block set with accessory clamps

Special Precautions:

• When adjusting the caliper, do not retract slide without carefully stoning the slide (if needed), to remove burrs or nicks which may damage the slide bushing.

• Whenever necessary to disassemble for adjustment, use care and cleanliness to assure no damage.

Tolerances:

Calipers with .001" graduations shall be within .001".

Calipers with .0005" graduations shall be within .0005".

Procedure:

• Verify that the identification marking of the caliper is distinct and in agreement with the Gage Management System Calibration Record.

• Visually examine it for obvious damage, other signs of mishandling and wear that may effect the accuracy or function.

• Clean contact faces with lint free cloth, dampened with cleaning solution.

• If any defects have been found that would affect accuracy or function of the caliper, discontinue calibration and refer to Gage Correction.

• Clean exterior surfaces.

• Remove slide assembly.

• Clean and oil slide.

• Clean and oil rack.

• Reassemble caliper.

• Check the rack for wear by pushing the slide to and fro. There should be no free movement between rack and slide. Adjust for wear if necessary by tightening fixed nut on barrel to a smooth tight (no shake) fit for the full length. If it is found that a smooth tight fit can not be achieved due to loose and tight areas, refer to Gage Correction.

• Check accuracy with gage block stacks having an accuracy not less than .00001". Standards chosen must test the caliper not only at complete turns of the indicator but also at intermediate positions. This is required as a check on the accuracy of the rack and pinion assembly driving the indicator.

• Arrange gage block combinations in such a manner so as to check the accuracy of the instrument in at least three positions of the total range. (zero, midway & full length).

• Record all appropriate readings in the Gage Management System.

• Label each instrument or case to identify the calibration date, next due date and person doing the calibration.

Gage Correction:

Any gage exceeding specified tolerances at any time during calibration shall be repaired and recalibrated, returned as is for restricted use, or scrapped.

All repair work shall be performed by approved repair personnel.

Another version of a work instruction that is worth mentioning separately is any instruction used to document and control your facility's accuracy tolerances or acceptance criteria. Every type of instrument used to confirm product can and may require a specific accuracy limit. Several approaches can be considered to establish these standards. Instrument OEMs often publish accuracy specifications in their product literature. Another resource that can be considered is the process of purchasing accuracy standards from organizations such as ASTM, ASME, JIS, or Global Engineering Documents, to name a few. You may hear some rather "left of center" suggestions from the "can't we just" crowd when it comes time to establish your facility's accuracy limits. I've heard things like "does the QS-9000 standard say that we have to go along with OEM specifications?" when I reply "no," then here it comes… "can't we just make our tolerances big enough that everything will always pass?" My answer to absurd suggestions like this lies in the concept and letter of a Quality System. Yes, a facility could develop an internal standard that states very clearly "all instruments must be accurate within ± 2.0mm" and the facility would (technically) satisfy element 4.11.2 which states: The facility must *"determine the measurements to be made and the accuracy required…"*

But what happens when your auditor or customer asks how this complies with the portion of 4.11.1 that requires: *"Inspection and test equipment shall be used in a manner that is consistent with the required measurement capability"*? You will be in no-man's land. Translation: If the accuracy tolerances of your calibrated gages are ±0.5mm (or ±.020"), then all of your customer part tolerances better be at least ±5.0mm (or ±.200")! Trust me, any "creative documentation" that is attempted when establishing internal accuracy tolerances will come back screaming for vengeance when you conduct your Measurement Systems Analysis (Gage R&R Studies), which we will discuss in Chapters 6, 7, and 8.

I would suggest looking no further than your gage software vendors to purchase a system within your software that has OEM, ASTM, ASME, etc. standards loaded within the software or available as an add-on. I have seen many of these, and Gage Tolerance Micropedia, for one, contains the essential features needed for most applications (*Figure 3-1*).

Another standard that you may get some interesting suggestions for involves documenting the calibration intervals for your facility. Once again, the QS-9000 stan-

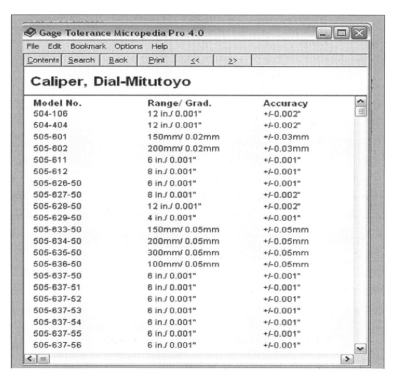

Figure 3-1: Accuracy tolerances by gage makers. Image courtesy Gage Tolerance Micropedia by Prompt Consultants, Inc.

dard offers plenty of rope for others to hang you with, as it appears to leave the acceptable calibration interval wide open in 4.11.2.

Identify inspection, measurement, and test equipment that can affect product quality and calibrate them at prescribed intervals, or prior to use, against certified equipment….

The "at prescribed intervals, or prior to use" almost guarantees that two common suggestions will surface when discussing calibration of equipment such as gage block sets used on the shop floor and/or pin gage sets that may have as many as 250 pins (or more) of (usually) .001" or .0005" increments. First of all, someone will ask: "can't we just calibrate the pin gages… <u>as a set</u>?" I've been asked this question several times, and it's usually asked in the context of an alternative to individually calibrating all 250 pin gages. I always ask the questioner to clarify EXACTLY what they mean by "as a set." I've never really had anybody who asked this question be able to clarify it. I think where they are going is to suggest calibrating every 5th or 10th pin gage, and then saying that "since those were good, we **considered** the others to be good." This is what they are leaning toward, but they'd like it to be someone else's idea so when it collapses like a house of cards during an audit they don't have to take responsibility. Therefore, my answer is "yes, we could do it as long as we make sure to calibrate our magic wand and pixie dust!" The QS-9000 standard says what is says, and it does not matter how far you twist it while you squint your eyes and plug your ears: it still means what it means.

After this is ruled out you'll most likely hear something along the lines of "well, can't we just calibrate each pin gage six-months after it's actually used, so we don't waste time calibrating pin gages that are rarely needed?" This may sound like a great idea in theory, and the powers that be swear it will save time and be easier. To prove that it's a roadmap to disaster, I'll play the role of your auditor and ask the questions any auditor worth his or her salt would ask about this calibration frequency. First, I'd I ask:

"How often are the pin gages calibrated?"

You'll reply (very confidently) with:

"Six-months after each initial use."

That seemed easy enough… but then I'm going to ask:

"How do you know when a pin gage is used?"

To which you'll have to reply:

"Our operators record it on this sheet."

Now you've added another controlled document to the process, the official "Pin Gage Initial Use Which Starts the Six-month Calibration Clock" sign out sheet (however titled). Then I'll ask:

"So, <u>every</u> time an operator uses a pin gage the first time after it is calibrated, he or she records it on the sheet?", and you'll reply:

"Yes, every time."

Then I'm going to ask:

"Since the different pin gages are calibrated at so many different times (due dates) how does the operator know they're using the pin gage <u>the first time</u> after a calibration?" And also,

"If an operator removes the .251" .252" and .253" pin gages from the set and walks over to his mill but only tries the .251" and .252" in a part bore, was the .253" considered <u>used</u> because it was removed from the set?"

Your guess is as good as mine as to what you'll reply. Whatever you reply, I'm going to ask a few operators to make sure that's <u>exactly</u> how they do it, <u>every time</u>. Then I'm going to ask:

"How do you label the pin gages that have been used. What if they're used today and need calibration in six months, but then they are used again next month and that date is mistakenly designated as the start of the six-month clock?" This question is a variation of a question I asked earlier (auditors are known to use this technique), so I should get the same answer.

No matter what you answer, I'm going to select about 15 to 20 random pin gages and ask to see documentation of when they <u>were initially used</u> (after the most recent calibration), and when they are due again. Then, I'll probably ask how this information is readily available to the operators to comply with QS-9000:

Remember, it is your responsibility to identify inspection measuring and test equipment with a suitable or approved identification record to show the calibration status. Keep in mind that whatever version of this pin gage frequency "can't we just" method you are using could easily result in over 100 separate due dates within a 250 pin set and these dates would never stop changing. Next, I would want to discuss the entire process with whoever is responsible with documenting these gage transactions because you have 20 to 30 people documenting dates on a controlled sheet, and not all of them are responsible for the overall integrity of the system. WOW! THIS IS SO EASY AND MAKES SO MUCH SENSE! I'll bet you've saved a bundle of time **"haven't you?"**

Sometimes you only need to play out these ludicrous ideas in your head as the Devil's advocate to determine that, like it or not, it's much safer to just calibrate the 250 pin gages today and set their due dates as six-months from now and forget about trying to find an easy way out. Any suggestion that begins with a variation of these three words: "can't we just…" is only suggesting what they hope can be done. Talking is cheap and easy. When you do what you say and prove it, then you're getting the job done.

When establishing calibration intervals for your own instruments, the only rule to follow is: The equipment should be calibrated frequently enough to detect unacceptable errors before suspect measurements are performed or out-of-tolerance parts are passed as acceptable. How is that for a roundabout answer? The only way you could possibly attain this would be to calibrate every instrument each time it is used. This is simply not practical. A balance must be achieved. We must find the middle of the road between "How do you know that interval is often enough," and "I really would sleep better if we calibrated that more often."

To assign the identical interval to all equipment is quick and easy and will gain management approval because management tends to like things neatly arranged in a small square box with no variables to understand or explain (that is, you explaining, them understanding). An established, across the board frequency that is written in stone usually guarantees only one thing: You'll over-calibrate some equipment while at the same time not keep tight enough control of other instruments.

I would recommend that you start with what you could realistically argue was an industry standard. In this context, we mean a rule of thumb. Most industrial facilities would recommend a six-month calibration frequency as a minimum. Even with this guidance you must remember that if you are borrowing from the experiences of others, you must compare their reality with yours. Are their conditions similar to yours? Is the skill level of their operators equal to the skill level of your operators? Are the tolerances and processes they are controlling tighter or more lenient than yours?

Some facilities send instruments to accredited outside laboratories and use the calibration interval recommended by the laboratory as their internal frequency. I wouldn't rely on this for a couple of reasons. First of all, it may be listed as 12-months, and if I were your auditor I would ask these questions:

"How do you know this interval is frequent enough?"

You'd most likely answer with a version of:

"That's what the lab recommended."

To which I would reply:

"So, when did they tour your facility and audit your processes to make this recommendation?"

You'd most likely have to admit that no such tour or recommendation had occurred, and then face the reality that you have relied on a standard laboratory calibration interval because your facility never considered that **your frequency requirements and the lab's standard interval are in two very separate worlds**. At this point, you might be asking the question: "Then how do I know six-months is frequent enough?" The answer: You don't! But the key to remember is the rule of thumb (six-month interval) represents what your customers, auditors, or other are used to seeing. If you fall into this category, the red flags will usually not fly. Keep in mind that we are aiming towards balance. If six-months does not seem comfortable enough for you, then start with 90-days for a couple of cycles and track the calibration results. If you're not finding out-of- calibration situations (which we'll discuss in Chapter 15), then bump the interval to 120-days and then eventually to six-months. It's your system and your protocol.

Protocol standards are just like swords: You may live by them, but you may also die by them.

Chapter 4
Getting Started

For every action, there is a reaction… For every cause, there is an effect…
For every beginning, there is an end…

Many facilities have a false hope that because they've had some sort of calibration system in place (even though it had no clear purpose, reason, or foundation), all that needs to be done is to make some minor adjustments and the system will "pass" without any problem. Albert Einstein once observed that the downfall of many systems and organizations occurred because people continually did the same things, yet expected different results. His thinking fits well with adapting to some basic concepts needed to get a decent measurement control system started. **Everything begins now:**

Sometimes in order to clean up a garage or attic, you must begin by making a mess. As we learned in Chapter 1, the QS-9000 standard can easily encompass all inspection, measuring, and test equipment that can or could affect product quality. The only way to even think about meeting this requirement begins with a complete inventory of **all** inspection, measurement, and test equipment in your facility. **By "all," I mean ALL**. Top to bottom, North to South, East to West, and Inside Out, Company owned or otherwise, everything needs to be inventoried and categorized. By doing this you can develop an **instrument master list**. I recommend separating the test equipment into like types or groups, such as but not limited to:

- Micrometers
- Calipers
- Dial Indicators
- Test Indicators
- Height gages

- Pin gages
- Plug gages
- Gage blocks
- Bore gages
- Etc.

I would further document everything that is known about each instrument, such as (but not limited to):

- Maker

- Model Number
- Resolution
- Range
- Manufactured date (if known)
- Replacement cost (if known)
- Type of application (depth micrometer, inside micrometer, etc.)
- Type of display (digital, dial, Vernier, etc.)
- Owner
- Normal storage location
- Normal location where the gage is used
- Etc.

I would use the following to break the instruments down into subgroups (groups within the groups):

- Resolution
- Range
- Type of display (digital caliper, dial caliper, Vernier caliper, etc.)
- Type of application (depth micrometer, inside micrometer, etc.).

Your list or matrix of instrument categories could easily encompass an entire chapter. Metrology is a game often won by those who pay the closest attention to the smallest of details. For this reason, I always recommend taking things at least one step further than (at the time) it seems like you need to go. This is for the four or five reasons that you can think of but, more importantly, the 10 to 15 reasons you can't think of. After you have a comprehensive list of instrument types, it's time to look at identification methods for each instrument, and accuracy levels for each gage type. The best approach is to divide, then conquer.

For each instrument to be properly controlled, and for this control to be documented, unique identification numbers are crucial. This is an easy task when dealing with hand-held precision measuring instruments such as micrometers, calipers, height gages, etc. Most precision measuring instrument manufacturers place unique identification numbers on the instruments they produce. Establishing control of other instruments, such as a complete set of pin gages, requires more ingenuity. With many makers of pin gage sets, the only unique number is assigned to the entire set and the unique pins are identified by the nominal sizes etched on each pin. As an example: If our shop had a .251" through .500" pin gage set that consisted of 250 pins in .001" increments, and the nominal size was labeled on each pin, we could use this as the unique identification number (*Figure 4-1*). The method would suffice as long as we did not have duplicate sized pins in our instrument inventory. The unique identification numbers could become: #PIN251, #PIN252, #PIN253, etc.

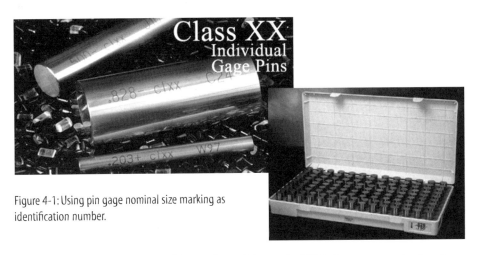

Figure 4-1: Using pin gage nominal size marking as identification number.

In the event there is an additional .251" through .500" pin gage set, the "can't we just" crowd will probably say "We just won't use the other set." Yeah, right. We didn't buy the additional set because we needed a paperweight! When we need to check an over-pin or between-pin measurement on a part, we'll need two pins of the same size. All you need to do is be able to differentiate from the two pins of the same nominal size. This is very easy. A technique I've used requires two different color paint markers. During initial calibration the end of each pin gage in set #A is colored with a green paint marker, and the pin gages in set #B are colored with a blue paint marker. After calibration each set can be looked at periodically in case any "touch up" is needed.

The unique identification numbers could become:

- #PIN251G
- #PIN252G
- #PIN253G
- #PIN251B
- #PIN252B
- #PIN253B

Like I've said before, it's not rocket science and I didn't invent it. It's as technical as it needs to be to work.

Another option for instruments that do not have unique identification from the maker is to label the identification number along with the calibration label. You do not have to look far to find labels formatted for this purpose. When you create the unique identification numbers for your instruments, there is one rule that will never let you down: incorporate the date because you can guarantee the number will not be duplicated by mistake later. If I were assigning numbers to 12 instruments today (6/8/05) I would use these numbers:

#RC060805-1, #RC060805-2. #RC060805-3, etc.

If I came in tomorrow (6/9/05) and labeled a few more, I could start with:

#RC060905-1, #RC060905-2, #RC060905-3, etc.

This may also have a hidden benefit that future circumstances will reveal. You will be amazed how handy it can sometimes be to know (in one glance) the date an instrument was initially controlled (*Figure 4-2*).

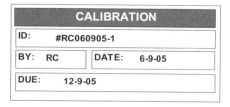

Figure 4-2: Unique identification numbers established using calibration labels.

Now is a good time to change our emphasis to instrument accuracy. Always remember that measurement is a science. Actually, without measurement there would be no other science. In any science there are fundamental rules. With any rules there will be exceptions. Both the rules, and why we can make exceptions, must be understood if we are to follow and hopefully master the science.

There's a fundamental, basic rule of metrology that states "you can never <u>expect</u> an instrument to perform better than its resolution." This rule will come into play as you establish accuracy tolerances for your facility's in-house calibration. We will use inch and Metric examples throughout this book. These dual examples will represent equivalent instrument examples, not exact conversions.

If we are using a .001" indicator (or 0.01mm indicator) we can never expect the indicator to measure better than ±.001" (or ±0.01mm). This can sometimes rattle cages because it seems like some instruments do perform better. Let's make that very clear: we can never <u>expect</u> the instrument to perform better than its resolution. If an indicator has a resolution of .001" (or 0.01mm), and we expected it to perform better than its resolution, we would be expecting the instrument to perform to an accuracy level of ±.000" (or ±0.00mm). This would be a perfect gage, which is a fictional item.

This fundamental rule throws several misconceptions out the nearest window. One of these occurs when an operator looks at a .001" indicator (or 0.01mm indicator) and says his or her part measures +0.0005" (or +0.005mm). Reading half-increments is a thing of the past (we'll discuss examples to back up this statement later in this book). So, one-increment is a good place to start with instrument accuracy.

If you look at accuracy levels published in the literature of instrument manufacturers, you'll see evidence of the one-increment rule displayed. The listed accuracy levels of the inch and Metric indicators fall within the one-increment rule. You'll find this pattern among several very popular measuring instruments. For example:

- Vernier micrometers .0001" resolution
- Vernier calipers .001" resolution
- Dial calipers .001" resolution
- Depth micrometers .001" resolution
- Height gages .001" resolution.

Like any rule, there are times this one cannot be applied. As resolutions get smaller and smaller, the laws of science and variation become more and more noticeable within every measurement taken. The thought of accepting an instrument error of ±.001" (or ±0.03mm) is not nearly as unbelievable as thinking the one-increment rule could apply at resolutions as small as .00005" (50 millionths) or 0.001mm (1 micron). This is just too small of an area of real estate to control the variables present. Usually you'll see three- to five-increments given up, as shown in *Figure 4-3*.

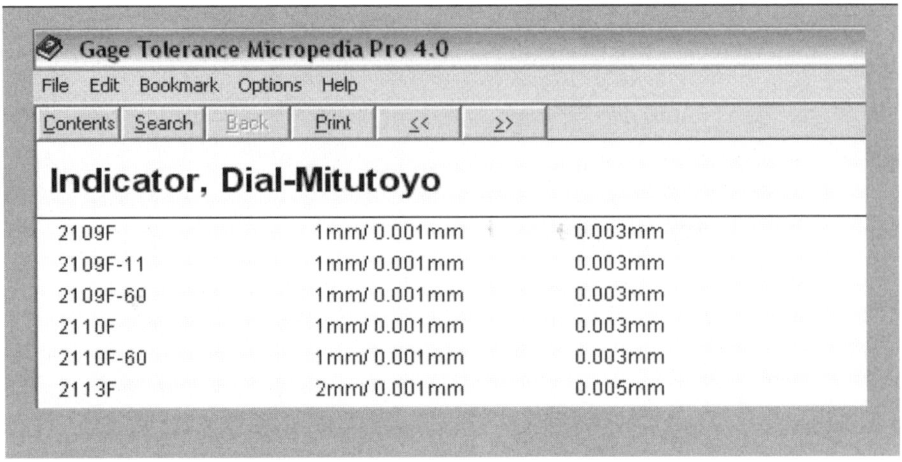

Indicator, Dial-Mitutoyo

2109F	1mm/ 0.001mm	0.003mm
2109F-11	1mm/ 0.001mm	0.003mm
2109F-60	1mm/ 0.001mm	0.003mm
2110F	1mm/ 0.001mm	0.003mm
2110F-60	1mm/ 0.001mm	0.003mm
2113F	2mm/ 0.001mm	0.005mm

Figure 4-3: Error allowed at calibration equals three- to five-increments. Image courtesy Gage Tolerance Micropedia by Prompt Consultants, Inc.

Now we can begin to understand what the QS-9000 standard is referring to when it states:

Determine the measurements to be made and the accuracy required and select the appropriate inspection, measuring and test equipment that is capable of the necessary accuracy and precision.

We can go back to the 10:1 rule and see how it plays out with some typical measurement scenarios. If we have a part tolerance of +.010"/ −.000", most would feel that we could use a .001" indicator because 10% of the .010" total tolerance is .001". But that might not be quite sensitive enough because the *accuracy and precision of* (or error from) the indicator *consumes* ± one increment. Indicator A could consistently measure +.001", while indicator B could measure −.001". Notice the QS-9000 standard does not say:

"A facility has to follow the 10:1 rule."

As your measurement system develops and Measurement Systems Analysis (MSA) studies (Gage R&R) are performed, any amount of "creative latitude" you may have placed into a measurement—such as relying on a 5:1 or even a 4:1 ratio—will bite you in a hurry. We'll discuss these situations and scenarios in later chapters, but for now just keep the 10:1 ratio in mind as a very fundamentally safe approach.

Another portion of the QS-9000 standard we should now give some thought to is the paragraph that states:

The supplier must identify all inspection measuring and test equipment that can effect product quality, and <u>calibrate and adjust them at prescribed intervals, or prior to use</u>, against certified equipment having a known valid relationship to international or national recognized standards.

The backseat drivers at your facility are almost certain to give you driving directions on this one. I'll discuss one concept right now that can be applied in many areas of the QS-9000 standard to make sure you don't make the wrong choice. I was once in a facility where instruments were documented in software for automatic calibration recall. This was a facility of over 300 employees with over 20 CNC lathes used in production environments, and all of the instruments were company owned. The instruments were not used as carefully as they should have been, and many were damaged and had to be exchanged before their scheduled six-month recalibration. As an example:

If 30 instruments were calibrated on 5/1/05 at lathe #C89, in six months (11/1/05) those 30 instruments will be due for recalibration. However, what usually happened over the next six months was that ten 10 of the gages (or so) would have to be replaced. These 10 instruments were calibrated before issue, so they were due six-months after the issue date.

Now, instead of having 30 gages due on 11/1/05, we would have only 20 gages due (the 20 that made it through the six-months without being replaced), while three replacement gages issued on 9/12/05 are due on 3/12/06; three gages issued on 10/19/05 are due on 4/19/06; and four gages issued on 10/22/05 are due on 4/22/06. This is displayed visually in *Figure 4-4*.

As 11/1/05 approached, I recommended recalibrating the 20 gages because they were due, and the auto recall feature in the software displayed calibration due dates and issued locations. The powers in charge said: "Wouldn't we save time running back-and-forth retrieving gages (on later dates) if we just recalibrated all the gages

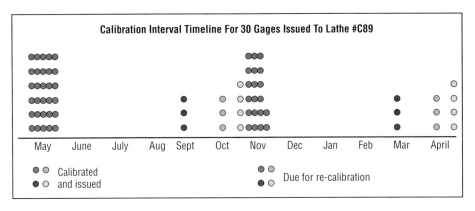

Figure 4-4: Calibration due dates after gages are reissued in mid-cycle.

that are now issued to lathe #C89 when the 20 are due on November 1st? After all, all of them would then be due on the same date again."

I asked if it would be saving time to recalibrate the four gages that were calibrated and issued on 10/22/05 (just over a week ago), and the gages calibrated and issued on 10/19/05. Their response was "Well of course not, they were just calibrated less than two-weeks ago."

So then I asked about the gages calibrated and issued on 9/12/05, and they said: "Yes, we should probably recalibrate those because they've been out there about six-weeks."

At this point, I had to ask "When <u>EXACTLY</u> is the cut-off date of how far away an instrument could be from its due date before over-calibrating would be 'saving time'?" Of course, nobody had an answer.

Always remember that the backseat drivers give directions based on a perfect world with no variables. A similar scenario played itself out in a movie from my years at Ft. Bragg. In "Platoon," the 1987 Academy Award Winner for Best Motion Picture, a soldier in the movie claims the Army should not send an infantry soldier out on missions when they were one week "short" (meaning they only had a week left in their 12-month tour of duty). Another soldier points out that, if this were the case, when a soldier had two-weeks left in his 12-month tour, he could consider himself a week "short" from the one-week out window (when he wouldn't be sent out on missions) and then he'd be in the same dilemma. He then stated that if the Army adopted this philosophy, they (the Army) would just keep moving the time backward a week before that and then a week before that, etc., and eventually "As soon as you jumped off the helicopter on your first day 'in country,' you'd be short already!" Calibrating 10 out of 30 (33%) instruments before their due dates to "save time" could easily follow the same pattern.

Bottom line: Getting started involves grasping only a few concepts; separating instruments into common groups and categories, establishing the accuracy specifications, the pros and cons of calibration intervals, and identifying each instrument by a unique number. This involves selecting the materials used to design and build your foundation of element 4.11.

Chapter 5
Common Outsourced Calibrations

Sometimes it's not what you know… It's who you know

A wise real-estate agent once gave me the best financial advice I've ever received. He said "shop for money like you'd shop for a house or a car." Many people go to great effort to find the house or car that meets their needs, then pull out all the stops during price negotiations. But they'll walk into one financial institution and sign their name on a loan without questioning a single item, because the bank is willing to lend them the money.

Not long after you begin developing an industrial measurement control system, you'll get to the time when outside source calibrations must be considered. Some items will need to be out-sourced because they will be used as internal standards—such as gage blocks identified as the "grandfather set" that will be used only for in-house calibration and measurement setup. Other instruments, such as granite surface plates, are usually calibrated by outside facilities because specialized equipment is needed to perform these calibrations or certifications. Regardless of <u>why</u> it's needed, the <u>who</u> must be decided upon.

Before choosing an outside facility to perform calibration services, you must have a solid understanding of what parameters need to be defined and met. Calibration and inspection laboratories become accredited to ISO-17025 through one of several accreditation bodies. A2LA, NVLAAP, and L-A-B, are the most common. The inspection labs usually display the certification along with a laboratory scope, or <u>Scope of Accreditation</u>. The scope is the document you need to review to find out the capabilities of each facility. How ISO-17025 accreditation works, in a nutshell, is that a laboratory is audited to the ISO-17025 standard and accredited to perform the inspections according to their scope. The certification itself is just a document stating the facility is accredited to perform measurements according to ISO-17025 standards. **The scope defines the measurements the lab is accredited to perform and the level of uncertainty with each accredited measurement.** We'll take a close look at the concept of uncertainty in Chapter 8, but for now just think of uncertainty as a "within" amount. If I can guess the actual time of day with an uncertainty of 30-seconds, then I can guess the time of day within 30-seconds of the actual time displayed by the atomic clock at N.I.S.T. If I estimate the time of day as 12:31:39 PM, the clock at N.I.S.T. will display a time somewhere between 12:31:09 PM and 12:32:09 PM. The accreditation concept is best demonstrated by two examples.

Let's assume I have a measurement facility and it's called Roadrunner-Coyote Inspections, Inc. I become accredited to ISO-17025 and my scope of accreditation lists one measurement: yardsticks with an uncertainty of ± 1.0000". This means that I can display an ISO-17025 accreditation logo on my literature and web-site, yet the only item I am accredited to calibrate is a yardstick to the nearest 1.0000".

Now let's look at another facility, Cincinnati Precision Instruments (CPI), which is accredited to ISO-17025 through A2LA. Their scope contains 15-pages of items to include: ring gage calibration with uncertainty of ± 10 microinches; indicator calibration with uncertainty of ± 18 microinches; and plug gage calibration with uncertainty of ± 9 microinches (*Figure 5-1*).

As can be seen in these examples, **displaying an accreditation logo really doesn't tell you anything about a lab. What the lab is actually accredited to measure, and to what level of uncertainty they can measure within, is what's important.** I would also recommend that you call prospective inspection laboratories and schedule a tour of their facility. Most are eager to accommodate this request. During this visit you can get some very good ideas that can be scaled down and implemented in your in-house facility.

As you begin selecting which outside labs you are comfortable with, you will then need to decide what level of calibration or certification you need to request. You often have a choice of different accuracy grades to choose from. Granite surface plates are calibrated using two specifications defined in Federal Specification GGG-P-463, which lists the following specifications (Courtesy of Tru-Stone Technologies, Waite Park, MN) for different grades of surface plates.

- Flatness. This specification means that all points on the surface of the plate will lie between two parallel planes separated by the flatness tolerance. Put another way, if you look at the surface plate from the edge, the difference between the lowest point on the plate and the highest point on the plate will be no more than the flatness specification. Manufacturers use three standard grades of flatness defined by the federal specification:

    o Laboratory grade AA: (40 + diagonal [in inches] of surface plate squared/25) × 0.000001 in.

American Association for Laboratory Accreditation

<u>SCOPE OF ACCREDITATION TO ISO/IEC 17025-1999
& ANSI/NCSL Z540-1-1994</u>

CINCINNATI PRECISION INSTRUMENTS
253 Circle Freeway Drive
Cincinnati, OH 45246
Sandy Frank   Phone: 513 874 2122

CALIBRATION

Valid until: December 31, 2006   Certificate Number: 1570.01

In recognition of the successful completion of the A2LA evaluation process, accreditation is granted to this laboratory to perform the following calibrations[1]:

IV. Dimensional

Parameter/Equipment	Range	Best Uncertainty[2,4](±)	Comments
Gage Blocks[3]	(0.01 to 4) in (4 to 20) in	(1.2 + 1.3L) μin (0.12 + 1.7L) μin	Federal comparator and gage blocks
Micrometers	(1 to 12) in (12 to 24) in (24 to 36) in	(110 + 3L) μin (220 + 3L) μin (350 + 3L) μin	GageMaker Mic-Trac Gage blocks
Calipers	(2 to 12) in (12 to 24) in (24 to 36) in (36 to 80) in	(110 + 4L) μin (170 + 4L) μin (250 + 0.4L) μin (300 + 0.4L) μin	GageMaker Mic-Trac Gage blocks Renishaw laser
2D Height Gages	(0 to 40) in	(98 + 1.5L) μin	Surface plate, gage blocks, height master, and reference bar
Bore Gages	—	(22 + 0.6R) μin	Ring gage and Indi-Check
Ring Gage - ID[3]	(0 to 3) in (3 to 20) in	10 μin (2.8 + 2L) μin	Zeiss ULM and gage blocks
Dial Indicators	(0 to 0.001) in (0.001 to 1) in (1 to 4) in	18 μin 82 μin 330 μin	Indi-Check Mic-Trac

Figure 5-1: Scope of ISO-17025 accreditation lists the (un)certainty of each measurement. Example courtesy of Cincinnati Precision Instruments.

- Inspection grade A: Laboratory Grade AA × 2
- Tool room grade B: Laboratory Grade AA × 4.

Depending on the manufacturer, this specification may be shown as total indicated reading (TIR), or as a plus/minus tolerance. Both mean the same thing. A surface plate with a TIR of 250 microinches is no different from a surface plate with a flatness of +/−125 microinches.

- Repeatability. The repeatability of surface plates is measured with a repeat measuring gage, sometimes called a Repeat-O-Meter, after the gage invented by Rudolf Rahn, a pioneer in granite surface plates. This instrument simulates placing a gage block and height gage on the surface plate. The gage is placed at the center of the surface plate and zeroed. As the repeat measuring meter is slid across the plate, the indicator will show any local deviation in the plate's flatness. Repeatability measurements are taken across the entire surface to ensure that there are no local peaks or valleys that fall outside the repeatability specification. This specification is much tighter than that of flatness, and also varies with the diagonal of the plate. For a plate with a diagonal of between 30- and 60-inches:
  - Laboratory grade AA: 45-microinches
  - Inspection grade A: 70-microinches
  - Tool room grade B: 120-microinches.

For example, to fully meet federal specifications, a laboratory grade AA 48 × 60 in. surface plate would require an overall flatness of within 280-microinches but must not have a localized variation of more than 45-microinches, as measured by a repeat measuring meter.

Cylindrical gages such as plug gages, pin gages, and ring gages are defined in different accuracy classes by the Gage Maker's Tolerance Chart. Depending on the nominal diameter of the cylindrical gage, the gage can deviate from its nominal and still fall within a given accuracy class (*Figure 5-2*). You must determine which accuracy class you are obtaining when you purchase new gages, and you must specify the accuracy class when you send existing gages to an outside lab for calibration.

Which accuracy class to choose depends on the application for which a specific instrument will be used. This is an area where management may want to come up with a blanket classification for all instruments. I don't recommend this because, in the long run, it causes more problems than it solves, and costs more money than it saves.

Cylindrical Plugs-General

Above	To and Including	XX	X	Y	Z	ZZ
0.020	0.825	0.00002	0.00004	0.00007	0.00010	0.00020
0.825	1.510	0.00003	0.00006	0.00009	0.00012	0.00024
1.510	2.510	0.00004	0.00008	0.00012	0.00016	0.00032
2.510	4.510	0.00005	0.00010	0.00015	0.00020	0.00040
4.510	6.510	0.000065	0.00013	0.00019	0.00025	0.00050
6.510	9.010	0.00008	0.00016	0.00024	0.00032	0.00064
9.010	12.010	0.00010	0.00020	0.00030	0.00040	0.00080

Above	To and including	XXXX	XXX
0.0009	0.825	0.000005	0.000010
0.825	1.510	0.000008	0.000015
1.510	2.510	0.000010	0.000020
2.510	4.510	0.000013	0.000025
4.510	6.510	0.000017	0.000033
6.510	9.010	0.000020	0.000040
9.010	12.260	0.000025	0.000050

Figure 5-2: Gage maker tolerances for cylindrical attribute gages. Image courtesy Gage Tolerance Micropedia by Prompt Consultants, Inc.

For example, let's imagine you have 50 different bore gage applications on your shop floor. You have multiple inside diameter part characteristics, and you use a bore gage and indicator combination that is zero-set using a ring gage. When dealing with a part tolerance of 34 +0.050mm, you must remain accurate within 0.005mm (0.050 × 10%) to meet the 10:1 rule. This will be very difficult because the manufactured accuracy of the bore gage and indicator may easily exceed 0.005mm. The last thing you'd want to do is add any more error within your zero setting. Your ring gage choices for a 34.000mm ring gages are:

1. Class XX, accuracy of 0.75-microns

2. Class X, accuracy of 1.5-microns

3. Class Y, accuracy of 2.25-microns
4. Class Z, accuracy of 3.0-microns.

For this application I would recommend using at least a Class XX ring gage to be as close to 34.000mm as you can in the zero setting.

If another part was listed as 50 +0.1mm, you could be a little less particular. You could "give up" 0.01mm of accuracy and still be within the 10:1 rule. In this case the 50.000mm ring gage choices are:

1. Class XX, accuracy of 1.0-microns
2. Class X, accuracy of 2.0-microns
3. Class Y, accuracy of 3.0-microns
4. Class Z, accuracy of 4.0-microns.

In this application you could most likely use a Class X or Class Y ring gage for the zero setting.

There will be times in industrial applications when higher accuracy classes are not even detectable from lower accuracy classes. I have seen this occur when dealing with the calibration of granite surface plates. Some facilities feel the plates located in the inspection area of quality labs should be lapped and calibrated to Grade AA because the measurements being made on these plates are so critical. Ironically, these same facilities often have very large plates located in these inspection areas, and the accuracy levels of granite plates are proportionate to their size.

If your parts are roughly one-cubic-foot in size, you could use a Grade B 18"× 18" granite plate, which is more accurate to measure from than a Grade AA 48" × 72" granite plate. The flatness accuracy difference between a Grade AA 48" × 72" granite plate and Grade B 48" × 72" granite plate is very seldom detectable when using measuring instruments common to industrial applications. The cost differences in these calibration grades are very substantial, as are the lapping fees when the plates are found to be worn outside of Grade AA specifications. You can wind up spending a lot of money on accuracy levels that are not needed in most applications.

Another oversight occurs because management may have the opinion that once a granite plate is calibrated within a certain accuracy grade, it will stay within this accuracy level for one-year. Guess again. How about a week or maybe even a day? The calibration certificate issued when a granite plate is calibrated certifies the plate measured within the flatness and repeatability specification on that given day. More specifically, the plate measured what it measured when it was cleaned thoroughly with a specific cleaner to remove dust, oil, or whatever else can lie between the actual granite surface and the part being measured. I've been in facilities that insist on spending thousands of dollars to have large granite plates lapped and calibrated within Grade A or Grade AA specifications, and yet they have no documented procedure regarding when and how the plates are cleaned and maintained. The calibration label or "sticker" on the side of the plate doesn't keep the plate within the flatness specification very well if the plate is never cleaned. Instructions on how to maintain the accuracy of granite plates are available from most granite plate makers (*Figure 5-3*).

Gage blocks also have specifications. For many years, GGG-G-15C was the accepted standard. This standard is currently being replaced by different ISO, ASME, and ANSI standards. The ASME specifications are referenced in *Figure 5-4*.

As in the applications involving plug gages, ring gages, and surface plates, your facility will have to decide which grade is the best choice for your applications. When making this choice, some hidden scenarios should be given some thought. When purchasing new gaging, you'll find it's often not much more expensive to purchase the higher grade. It may seem like a wise choice to spend two- or three-percent more money for twice the accuracy. Be advised that, due to normal use, attribute gages will not retain these accuracy levels forever. If you purchase the higher grade it may seem like a wise choice until the gage block set is sent out to an outside laboratory for recalibration and 25% of the gage blocks within the set are not within the purchased grade. Do you remember this QS-9000 requirement discussed in Chapter 1?

Access and document the validity of previous inspection and test results when inspection, measuring or test equipment is found to be out of calibration.

Given this requirement, buying the higher grade for just a few percent more might not be a wise choice in the long run. Granite plates are the same way. You might be able to swing a nice discount by purchasing a fleet of Grade AA plates, but they won't always be Grade AA. At some point you'll have to explain how and why you know they are adequate, even though the calibration certificate states: OUT OF TOLERANCE. It's a simple case of possession being nine-tenths of the law, or perception being as important as truth. It's usually easier to start with a lower, but more realistic accuracy grade, that the instruments have a better chance of maintaining than starting with a higher grade and explaining (with documentation) how you know they're

Instructions for Care of Granite Surface Plates

1. **Cleaning and Moisture:** Plates should be thoroughly cleaned and allowed adequate time to dry before testing for tolerance. Water based cleansers that have not dried will cause iron parts to rust if they are left in contact with the wet surface for an extended period of time. It is recommended that plates undergo drying time in a room with less than 50 percent relative humidity. Temperature and dirt have a direct correlation with measurement accuracy. Personal cleanliness will aid in eliminating one source of contamination.
2. **Temperature Soaking Time:** Before granite surface plates are measured for work surface flatness, the granite should remain in the calibration area until it has reached room temperature, which may require 2 to 3 days. Large plates require more soak-out time than smaller ones.
3. **Scratches and Nicks**: Whenever scratches and nicks appear on granite plates, the resulting rough edges should be removed with a flat granite dressing plate. Any bump that shatters the surface raises fractured material at the rim of the crater.
4. **Rotation of Plates:** When a specific work surface area receives prolonged usage, it is suggested that the plate and stand be rotated 180 degrees on a periodic basis to increase the wear life of the plate. The production of a contour map during calibration is particularly helpful in locating the parts of the plate that should be given the most use. This can be accomplished by requesting a long form certification when ordering the new surface plate or when the plate is being sent in for recalibration.
5. **Periodic Calibration:** Periodic calibration of granite surface plates is recommended to determine resurfacing or replacement needs. The interval between calibrations will vary with the grade of plate and the wear resistance of the granite. TRU-STONE CONFORMS TO ISO 9001 and ISO/IEC 17025 CERTIFICATION REQUIREMENTS. Our calibration certificates are NIST Traceable. Frequent monitoring of the work surface by scanning it with the repeat reading gage is desirable. When these results differ from those marked on the replaceable sticker, you should recalibrate the plate. In addition to measuring the overall accuracy of a surface, smaller areas can be checked for localized variations often missed by the calibrating methods. Remember precision measurements are only as accurate as the measuring tools used.
6. **Torque on THREADED Inserts:** Do not exceed the following maximum torque values when using a torque wrench to limit distorting the work surface and pulling the insert. The following torque values are the maximum level permissible by the Federal Specification GGG-P-463c.

PERMISSIBLE TORQUE CLAMPING ON THREDED INSERTS

Thread Size	Torque
.250 inch	7 ft. lbs.
.3125 inch	15 ft. lbs.
.375 inch	20 ft. lbs.
.500 inch	25 ft. lbs.

7. **Clamping Ledges on Grade AA Surface Plates:** There is danger of distorting the work surface flatness beyond tolerance when a heavy item rests on the ledge or an item is clamped to the ledge. Ledges are not only expensive, but a great cause of inaccuracy as well. Experimentation and research reveal that no-ledge plates retain their accuracy better than ledged plates.
8. **Supports:** There are working and loading conditions where the standard three point supports are not satisfactory. These cases should be individually engineered. When four or more

1101 Prosper Drive • P. O. Box 430 • Waite Park, MN 56387 • 1-800-959-0517 • PH: 320-251-7171 • Fax: 320-259-5073
http://www.tru-stone.com

Figure 5-3: Instructions for care of granite surface plates. Image courtesy of Tru-Stone Technologies.

supports are used, shims or adjusting screws are necessary for proper support. The supports could be spotted under the loading points and set to approximately equal the loading. Sometimes the work surface flatness can be improved by shifting support positions. Fulcrum, air and hydraulic supports are available. Whenever nonstandard supports are used, the surface plate shall be calibrated at the site for compliance to the flatness tolerance.

9. **Care:**
   - Utilize the full surface of a plate so the wear is distributed and not concentrated in one area.
   - The surface plate should not be overloaded.
   - Use extreme care in moving the item being measured and the gages being used.
   - Place on the surface ONLY what is required.
   - Particularly avoid heavy contact with the edges.
   - Don't leave metal objects on the surface longer than necessary.
   - Clean the surface before and after use.

> Remember that the condition of this accurate plane
> is an integral factor in the measurement being made.

NOTE:
Surface plate cleaner can be purchased through your local Tru-Stone distributor or directly from Tru-Stone Technologies. Call for pricing and delivery information.

When the need for recalibration or rework of your surface plates and precision granite accessories arises, contact a Tru-Stone representative or call us directly for more details on restoring your inspection item to a "like new" condition. This service comes with new certification and plate labels.

One advantage of having your granite inspection equipment recalibrated by a manufacturer is the manufacturer's ability to take the time required to ensure the proper repeatability and overall shape of the inspection surface. Tru-Stone allows the item to normalize overnight prior to taking the final readings to indicate whether it is a good enough quality to certify and return to you.

Should you have any questions, please contact our customer service representatives at 320/251-7171 or fax us at 320/259-5073.

Important Notice

Any streaks of color in the granite are not defects. These are created by the molten lava mixing with minerals prior to evolving into the granite you see today. Black streaks or spots are the result of a magnetite (black iron oxide) concentration. White streaks or spots are areas where the granite is lacking magnetite. The levels of magnetite in granite vary considerably, however the color of the granite in no way affects the functionality or quality of it.

Our products are UNCONDITIONALLY Guaranteed

Please refer to the Federal Specification GGG-P-463c, which is followed by NIST (National Institute of Standards & Technology) for Granite Surface Plates.
- "3.7 Seams or Color Streaks: *Seams are cause for rejection. Color streaks have no affect on the serviceability of the granite.*"
- "4.5.8 Seams or Color Streaks: *Test for a seam is to wet the smooth surface of the granite where the color streak appears; then dry it off. If the streak remains wet or damp, it is a seam.*"

Figure 5-3 (continued): Instructions for care of granite surface plates. Image courtesy of Tru-Stone Technologies.

Nominal Length Range l_n in mm	Calibration Grade K		Grade 00		Grade 0		Grade AS-1		Grade AS-2	
	Limit Deviations of Length at any Point From Nominal Length $\pm l_e$ µm	Tolerance for the Variation In Length t_v µm	Limit Deviations of Length at any Point From Nominal Length $\pm l_e$ µm	Tolerance for the Variation In Length t_v µm	Limit Deviations of Length at any Point From Nominal Length $\pm l_e$ µm	Tolerance for the Variation In Length t_v µm	Limit Deviations of Length at any Point From Nominal Length $\pm l_e$ µm	Tolerance for the Variation In Length t_v µm	Limit Deviations of Length at any Point From Nominal Length $\pm l_e$ µm	Tolerance for the Variation In Length t_v µm
$l_n \leq 0.5$	0.30	0.05	0.10	0.05	0.14	0.10	0.30	0.16	0.60	0.30
$0.5 < l_n \leq 10$	0.20	0.05	0.07	0.05	0.12	0.10	0.20	0.16	0.45	0.30
$10 < l_n \leq 25$	0.30	0.05	0.07	0.05	0.14	0.10	0.30	0.16	0.60	0.30
$25 < l_n \leq 50$	0.40	0.06	0.10	0.06	0.20	0.10	0.40	0.18	0.80	0.30
$50 < l_n \leq 75$	0.50	0.06	0.12	0.07	0.25	0.12	0.50	0.18	1.00	0.35
$75 < l_n \leq 100$	0.60	0.07	0.15	0.07	0.30	0.12	0.60	0.20	1.20	0.35
$100 < l_n \leq 150$	0.80	0.08	0.20	0.08	0.40	0.14	0.80	0.20	1.60	0.40
$150 < l_n \leq 200$	1.00	0.09	0.25	0.09	0.50	0.16	1.00	0.25	2.00	0.40
$200 < l_n \leq 250$	1.20	0.10	0.30	0.10	0.60	0.16	1.20	0.25	2.40	0.45
$250 < l_n \leq 300$	1.40	0.10	0.35	0.10	0.70	0.18	1.40	0.25	2.80	0.50
$300 < l_n \leq 400$	1.80	0.12	0.45	0.12	0.90	0.20	1.60	0.30	3.60	0.50
$400 < l_n \leq 500$	2.20	0.14	0.50	0.14	1.10	0.25	2.20	0.35	4.40	0.60
$500 < l_n \leq 600$	2.60	0.16	0.65	0.16	1.30	0.25	2.60	0.40	5.00	0.70
$600 < l_n \leq 700$	3.00	0.18	0.75	0.18	1.50	0.30	3.00	0.45	6.00	0.70
$700 < l_n \leq 800$	3.40	0.20	0.85	0.20	1.70	0.30	3.40	0.50	6.50	0.80
$800 < l_n \leq 900$	3.80	0.20	0.95	0.20	1.90	0.35	3.80	0.50	7.50	0.90
$900 < l_n \leq 1000$	4.20	0.25	1.00	0.25	2.00	0.40	4.20	0.60	8.00	1.00
Nominal Length Range l_n in inches	Calibration Grade K		Grade 00		Grade 0		Grade AS-1		Grade AS-2	
	Limit Deviations of Length at any Point From Nominal Length $\pm l_e$ µin.	Tolerance for the Variation In Length t_v µin.	Limit Deviations of Length at any Point From Nominal Length $\pm l_e$ µin.	Tolerance for the Variation In Length t_v µin.	Limit Deviations of Length at any Point From Nominal Length $\pm l_e$ µin.	Tolerance for the Variation In Length t_v µin.	Limit Deviations of Length at any Point From Nominal Length $\pm l_e$ µin.	Tolerance for the Variation In Length t_v µin.	Limit Deviations of Length at any Point From Nominal Length $\pm l_e$ µin.	Tolerance for the Variation In Length t_v µin.
$l_n \leq 0.05$	12	2	4	2	6	4	12	6	24	12
$0.05 < l_n \leq 0.4$	10	2	3	2	5	4	8	6	18	12
$0.45 < l_n \leq 1$	12	2	3	2	6	4	12	6	24	12
$1 < l_n \leq 2$	16	2	4	2	8	4	16	6	32	12
$2 < l_n \leq 3$	20	2	5	3	10	4	20	6	40	14
$3 < l_n \leq 4$	24	3	6	3	12	5	24	8	48	14
$4 < l_n \leq 5$	32	3	8	3	16	5	32	8	64	16
$5 < l_n \leq 6$	32	3	8	3	16	5	32	8	64	16
$6 < l_n \leq 7$	40	4	10	4	20	6	40	10	80	16
$7 < l_n \leq 8$	40	4	10	4	20	6	40	10	80	16
$8 < l_n \leq 10$	48	4	12	4	24	6	48	10	104	18
$10 < l_n \leq 12$	56	4	14	4	28	7	56	10	112	20
$12 < l_n \leq 16$	72	5	18	5	36	8	72	12	144	20
$16 < l_n \leq 20$	88	6	20	6	44	10	88	14	176	24
$20 < l_n \leq 24$	104	6	25	6	52	10	104	16	200	28
$24 < l_n \leq 28$	120	7	30	7	60	12	120	18	240	28
$28 < l_n \leq 32$	136	8	34	8	68	12	136	20	260	32
$32 < l_n \leq 36$	152	8	38	8	76	14	152	20	300	36
$36 < l_n \leq 40$	160	10	40	10	80	16	168	24	320	40

NOTE: Grade K is direct measurement by interferometer.

Mitutoyo

Figure 5-4: ASME gage block tolerances. Image courtesy of Mitutoyo America.

still okay, even when they're out of tolerance. Trust me… it's complicated just writing about these scenarios, let alone being in the actual situation where 25% of your gage blocks (used to calibrate 90% of your instruments) are declared out of tolerance.

The general consensus of management is that you simply call a lab and ask them to come calibrate a piece of equipment. Learning a few simple parameters about these items can lower costs, save a facility a lot of money in the long run, and assure the services requested are the services you receive—no more and no less.

Chapter 6
Scheduling Outsourced Calibrations: Tips and Tricks That Can Save Time and Money

Working smart beats working hard… every time.

When scheduling and facilitating outside source calibrations (especially on-site services), it is crucial to remember the 5Ps: **Prior Planning Prevents Poor Performance.** This preparation begins before as the Purchase Order (P.O.) is typed up and faxed or e-mailed to the outside calibration source.

The purchase order is the contract between the facility needing a calibration or inspection service and the outside source calibration provider. If a P.O. states only "24 surface plates need to be calibrated" it can cause gaps in the systematic loop that should run between your facility's quality department, your facility's purchasing department, and the outside source calibration provider. If you have 24 surface plates that need to be calibrated and relapped (if necessary) within Grade B specifications, then your P.O. should state: "24 surface plates need to be calibrated and relapped (if necessary) within Grade B specifications." It also wouldn't hurt to add "complete with long form certifications with data listing as-found and as-left conditions" and "inspections must be accredited to ISO-17025" along with a "no later than" date after the on-site dates have been agreed upon.

The more detailed the P.O., the more assurance your facility has against "misunderstandings." Case in point: I was involved in scheduling an on-site calibration at a facility that needed 40 granite surface plates calibrated and relapped on-site. I scheduled the calibration with a very large "Nationwide" type calibration provider that was ISO-17025 accredited to perform the calibration on-site. I clearly stated three times during phone calls that we had 40 granite plates, and I suspected at least 20 would need to be relapped to measure within Grade B specifications. When the technician arrived on the scheduled date I found out the very large calibration provider actually used subcontracted granite calibration providers. This was fine with me, but there was a small communication problem within the equation. The subcontractor was told we had 20 Granite plates as opposed to 40. Now the subcontractor was in a tight spot because this was a Monday and he had scheduled his direct customers for all four-days the remainder of the week. It was somewhat of a messy situation. I believed he had been given incorrect information and I understood why he didn't want to delay his own customers over an issue that was not at all his fault with a sub-contracted job. He wound up getting what he could calibrated during the first day (about 25 plates), and we rescheduled (with another facility) to have the other 15 plates calibrated a few weeks later. What I learned during that experience was priceless.

Now when I have outside source calibration facilities asking for business, I explain to them that the granite on-site is what I use as a barometer. If they'd like to perform services for us, we'll try the granite on-site first, and use this experience to form our impression of their services. We will use this impression to decide on the future outside source calibration services that we'll need them to provide. This nearly guarantees no "misunderstandings" like the one that the subcontracted granite plate calibration technician and I were blessed with. This has turned out to be one of the best strategic moves I've ever made with outside source calibration providers. The labs earn business the old fashion way, by working for their reputation. If that's not how they think it ought to be, they're probably in the wrong profession.

Another very smart option to consider, after your system has developed and outside source calibration needs are established, is to **work with your chosen calibration provider to create an annual calibration plan** or forecast. Many outside source calibration providers will offer a discount of as much as 5% to 10% if all of your calibrations can be arranged for the year. **It becomes an open P.O. and the calibrations can be scheduled as needed throughout the year wi**th one phone call or e-mail. This saves additional money by eliminating many meetings and discussions throughout the year. You'd be surprised how many facilities I've seen that—even though a calibration was performed last year and needs to be performed again this year in order to continue complying with QS-9000 or ISO—argue among themselves, and management needs to be "talked into" signing the approval. With an annual contract and purchase order, it becomes very easy for a facility to schedule a 15- to 30-minute meeting with all required members of management, review the quote for all calibrations needed for the year, and sign-off and approve everything for the next 12-months. This saves an incredible amount of time.

If these calibrations are scheduled individually, every person that's in the loop of request/approval at your facility could easily spend 15-minutes processing his or her portion of the transaction. If there are 5 people in this loop and there are 10 separate calibrations that are performed by outside sources during any given year, this adds up to:

5 people × 15 minutes × 10 calibrations = 750 minutes (12.5 hours)

The result is some very expensive arguing that often winds up right back where you started… a year ago.

Another area where time and effort can add up to a costly bottom line is in the area of instrument repair. If most of the controlled instruments used at your facility are company owned, multiple operators and/or shifts probably share them. The nature of this beast will result in instruments that are damaged and need repair. This seems like a simple task: Find a repair facility or send the instruments to the repair department of the OEM. Quite simple… and quite costly. The cost is not really in the repair, arguing about how to best do it equates to pouring money down the drain. Costs can be held to a minimum by not wasting a lot of time (along the same lines as the above example that wastes 12.5 hours) trying to save perhaps $50. Let me explain a scenario that played itself out at a facility I was working with on the West Coast.

This facility had several production lines with an adhesion process that permanently bonded paper onto a steel plate. The temperature of the molding plate used in the process was controlled with an in-line thermometer, and the customer control plan required that a manual temperature measurement be taken on each plate type every two-hours. This facility used a very accurate and fast reading thermometer and a wand-type probe that cost around $350 ($200 for the fast response wand and $150 for the hand-held digital display unit). During eight months of normal use, the wand developed a short in the wiring (inside the wand). We could determine this because the display unit would measure correctly using another wand from one of the other processes. The operators would now have to walk to another process and borrow an instrument until their instrument could be repaired or replaced.

When the gage control technician submitted a requisition form to purchase a new wand, at a cost of $200, management asked what they always ask: "Can't we just get the existing one repaired?" What they failed to understand was that electrical type instruments are often manufactured at very large, world-class production line facilities. A new wand would be delivered two to three days after a purchase order was faxed or e-mailed to this provider. When you send an instrument like this to the OEM for repair, the repair department is most likely not capable of "kicking out" an instrument every 30- to 60-seconds like the production lines where the instruments are made can do. The instrument would have to be sent to the repair department where the damage could be assessed. Even though the purchase order clearly states "repair as needed," the repair facility will most likely fax an estimate and approval form before any repair will be attempted, and sometimes before spare parts are ordered. This will most likely be faxed to your purchasing department, and may or may not be at the top of their list. If you're lucky, the repair will be only about half the cost of a new wand, or $100. You see, it's kind of like going to the doctor, even if he pats you on the back and tells you to get more sleep, the bill is still $60. After approval, the repair is almost guaranteed to take an additional three- to four-weeks and will probably be shipped by camel.

Do you see what's going on here? For starters, the instrument was damaged after eight months of normal use (we could probably stand to keep a spare on hand). We invested time, effort, and frustration for five- to six-weeks, during which time our operators have walked to another process four times every shift to borrow a thermometer and another four times to return it. Now, if the operators are human, they probably strike up a conversation once or twice while they are over there. How much lost production time does it take before we've wasted $100? Not much. What we should have done was get a purchase order out to the provider as soon as possible to get a (new) replacement en route, and then use the repaired instrument as a spare when it finally found its way back to our facility. Losing time is like losing interest on a savings account, you can always put the money back in the bank account later, but the lost interest is lost forever.

Confusion and misunderstandings are additional ways to waste money. There are very effective ways to limit miscommunications, and they can be found by looking at some other precision oriented professions. When working with industrial precision measurement, it is very common to encounter letter/number combinations. This often occurs when referencing control numbers on customer's drawings as well as instrument identification numbers. In a noisy, production driven environment, asking someone to find #1D4662 can often sound like #1C4662 or #1B4662. I've been in situations on the shop floor where a specific drawing or customer standard was needed to get a machine set up to run some new model test parts. We ask for the engineering department to bring us the most current version of the drawing for the #MBJ 3rd plates, but we received the drawing for the #MDJ 3rd plates by mistake. Every minute spent running the new model parts was a minute that current, mass production parts, would not be running. <u>We wasted an extra 20-minutes because of a miscommunication on one letter!</u>

With effective communication being a common "weakest link" in a chain, it is recommended that you consider communicating in a more effective manner. For many, many, years, pilots (as well as units of the military and emergency personnel) have used a method called the phonetic alphabet. Using the phonetic alphabet eliminates any confusion of "did he say one-six-E, one-six-D, or one-six-B?" Each of the 26-letters of the alphabet is designated by a specific word that starts with the letter:

A = Alpha	N = November
B = Bravo	O = Oscar
C = Charlie	P = Papa
D = Delta	Q = Quebec
E = Echo	R = Romeo
F = Foxtrot	S = Sierra
G = Golf	T = Tango
H = Hotel	U = Uniform
I = India	V = Victor
J = Juliet	W = Whiskey
K = Kilo	X = X-ray
L = Lima (pronounced Lee-ma)	Y = Yankee
M = Mike	Z = Zulu

This method takes all the guesswork out of the numbers and letters. Had we used this on the machine setup I mentioned earlier, we could have asked engineering for the most current version of the drawing for the "Mike-Bravo-Juliet" 3rd plates. We would have received the drawing we needed and by using only three words, we would have saved 20-minutes of production time. Fire prevention trumps fire fighting… every time!

Another missed opportunity occurs when an industrial calibration technician receives an instrument returning from an outside source calibration, removes the calibration certificate, and places the certification in a binder until he or she needs to show it to an auditor or customer. What was the missed opportunity? Outside source calibration certificates can be used as sacred sources of knowledge and understanding. These certificates document how the equipment you use to calibrate gages and confirm products has been calibrated and confirmed. I recommend studying every line, term, standard (both protocol and artifact), and entry of data listed on the certification. If you don't understand what it means, pick up the phone and find out. You'll be amazed at what valuable "cross-talk" you can pick up, not to mention that your outside source calibration facilities enjoy informed and educated customers because their best publicity is a customer tooting their horn for them.

Knowledge is power, get it every chance you have!

Chapter 7
Learning, Understanding, and Conquering Thermal Effects in Industrial Measurement

> *"Energy can neither be created nor destroyed,*
> *it can only be transferred."* —Isaac Newton

When starting an industrial measurement control system, thermal effects in measurement and how to accommodate them can make you feel like you are standing in front of a field of haystacks looking for a needle. There's an anthem among many accredited outside source calibration facilities: "The Lab with the best thermometer wins!" Trust me when I say that, like everything else in precision measurement, it can be as simple as counting beads on a string, combined with some Jr. High level mathematics. Before we get started, we'll review what the QS-9000 standard says about environmental conditions and your measurements.

> *"The supplier shall… ensure the environmental conditions are suitable for the calibrations, inspections, measurements, and tests being carried out."*

This becomes a problem if you cannot literally give a class on exactly what this statement does (and does not) mean. Then you are at the mercy of your auditors, your customers, and the misperceptions that have been assumed true for quite some time. Unfortunately, for most who are tasked to set up an in-house measurement control system, the most accessible people to ask about thermal effects are our friends at the outside source calibration labs whom we depend on several times per year. This can be unfortunate because, even though these people are extremely knowledgeable about this subject, they deal with these issues in situations where objects must be controlled and measured in increments as small as five-millionths of an inch. These applications require ambient (air) temperature to remain around 68°F ±2° and gage temperatures around 68°F ±0.5° (during the 1800s, 68°F, equal to 20°C, was set as the "standard" temperature for calibrating gages). Most industrial inspection facilities don't measure parts in this "micro-world," but more in the "real-world." Remember from Chapter 3; if you are borrowing from the experiences of others, you must compare their reality with yours.

There are two thermal effects in industrial measurements that must be learned and understood:

1. Thermal expansion.

2. Thermal transfer.

To begin with, we should look at the very basic equation used to calculate thermal expansion (*Figure 7-1*).

Change in length = Original length × Coefficient of thermal expansion × Change in temperature from 20°C.

$$\Delta L = L \text{ original} \times CTE \times \Delta T$$

Figure 7-1: Equation for calculating thermal expansion.

We should note one thing about the phrase "change in temperature from 20°C." The standard measurement temperature of 20°C (or 68°F) is normally as cool of an environment as you'll be measuring in. However, thermal effects are constant, whether the object is expanding from increasing temperature or contracting from decreasing temperature.

Every man-made material on the face of the Earth has a coefficient of thermal expansion (CTE), and ASME Y14.5 – 1994 (The current standard which governs the Geometric Dimensioning and Tolerancing of your customer's drawings), states

"*Unless otherwise specified, all dimensions are applicable at 20°C (68°F). Compensation may be made for measurements taken at other temperatures.*"

If we wanted to calculate the thermal expansion of a 101.600mm steel piece or gage block that was being measured in a 76°F environment, we'd use the CTE for steel (0.0000115 as seen in *Figure 7-2*). 76°F represents a realistic shop temperature, for at least three-months out of the year.

Change in Length = 101.600mm × 0.0000115 × 4.44°C (Difference from 76°F and 68°F)

Change in Length = 0.00519mm (or 5.19 microns)

Expanded Length = 101.60519mm

If you prefer using just inch measurements, we could calculate the expansion of a

4" steel piece or gage block another way. If we use inch and Fahrenheit, we can simply multiply the CTE (of 0.0000115) by 0.556 (the conversion factor for converting CTEs from Celsius to Fahrenheit).

CTE = 0.0000115 × 0.556

CTE = 0.0000064

Change in Length = 4.00000" × 0.0000064 × 8°F (Difference from 76°F and 68°F)

Change in Length = .000205" (or 2 tenths)

Expanded Length = 4.0002".

If you really want to learn the concept forwards and backwards, I suggest confirming your calculations using the 1" = 25.4mm conversion factor. A steel piece that is 101.600mm long is the same length as a 4.0000" steel part. Our metric conversion estimated the thermal expansion on the 101.600mm part (temperature difference from 76°F and 68°F) of 0.00519mm while our 4.0000" part expanded .000205" and, guess what...

0.0052 / 25.4 = .0002 and .000205 × 25.4 = 0.0052.

Now we are right in the middle of no-man's land. If we're calibrating a caliper with an OEM accuracy tolerance of ±.001", the caliper is not sensitive enough (with a

The thermal expansion coefficient of CERA gage blocks is quite similar to that of steel.				
Property/Material	CERA Block (ZrO$_2$)	Steel	Carbide (WC-Co)	Silicon nitride (Si$_3$N$_4$)
Hardness (HV)	1350	800	1650	1500
Thermal expansion coefficient (10^{-6}/K)	10±1	11.5±1	5	2
Flexural strength (three-point bending) (kgf/mm^2)	130	200	200	60
Fracture toughness Kic (MN/N$^{1.5}$)	7	>20	12	6.5
Young's modulus (x10^4 kgf/mm^2)	2.1	2.1	6.3	2.9
Poisson's ratio	0.3	0.3	0.2	0.3
Specific gravity	6.0	7.8	14.8	3.2
Thermal conductivity (cal/cm • sec°C)	0.007	0.13	0.19	0.04

Figure 7-2: Coefficient of thermal expansion for steel. Example courtesy of Mitutoyo America.

resolution of .001") to detect the expansion of the 4" gage block. The same concept would apply if we were measuring a 4" workpiece which, because of an open tolerance, could be inspected with a caliper.

Fabulous! At 76°F we have not violated the mighty Element 4.11, but wait... what if we are inspecting a 4" workpiece, using a .0001" resolution micrometer, and the tolerance of the piece is plus .0005" minus nothing. We measure the diameter 15 different times and get 4.0001" (12 times) and 4.0000" in three narrow places. Now, our Ace will surely get trumped. What consistently measures 4.0001" at 76°F is actually 3.9999" (.0001" under specification) at 68°F. We should take notice at this time of the fact we've just discussed and "figured out" two very real thermal effect situations, and we used nothing more than a $5 calculator.

Some people I've dealt with over the years insist: "Thermal effects aren't real because parts expand and gages expand, so it really doesn't matter," or "Do we really have to recognize thermal effects?" (It's a law of science. I suppose we could countermeasure a problem of damage from parts being dropped by writing an approved work instruction stating "*Gravity does not exist in our facility*.")

The approach used in the above paragraph is not correct. However, in some cases, understanding how the effects work can give you the opportunity to cheat the physics that make them occur. For example: If we were reworking a step height on several

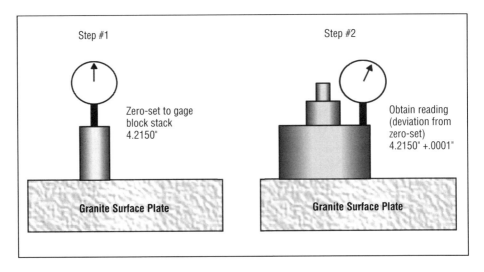

Figure 7-3: The part (on the right) is measured using the deviation from the master setting (gage block stack on left).

pieces in our 76°F shop, and because of the schedule we needed to inspect them at the machine and not across the shop in the lab, we could set up a gage block stack at the specification nominal (we'll say 4.2150") and use a .0001" dial indicator, set to the gage blocks (*Figure 7-3*). We would then be able to measure the deviation of each piece to confirm it's within the (+.0005"/−.0000") tolerance.

Now we have a dilemma. We know the thermal expansion in our 76°F shop on a 4" part causes the part to "grow" two-tenths (meaning two-ten-thousandths of an inch), which may lead us to believe our (4.2151") part will measure too small at 68°F. Understanding the simple science and setup of how and when thermal effects occur allows us to rest easy with this application. There is a thermal effect within this application, but the expansion is constant between the gage block stack and the part. If the part measures .0001" larger than the gage block stack in the 76° F shop environment, then it will also measure .0001" larger than the gage block stack when the gage blocks and the part are allowed to stabilize or "soak" to 68°F. This concept holds as long as the gage block stack remained in the 76°F shop environment long enough prior to setup to stabilize to the part temperature.

Thinking about our earlier example using a .0001" resolution micrometer; we could obtain separate 1", 2", and 3" gage blocks and have them calibrated annually with our 81-block set. These additional gage blocks can be stored (with care) in the shop to use as **controlled** micrometer setting masters. Now the zero setting of our micrometer is thermally stable to the parts being measured.

Before too long you will ask (or will be asked) the question: How long do we need to let instruments "soak" (the not-so-technical term for the thermal stabilization process) in our in-house laboratory environment before we can measure, inspect, or calibrate them? There are several quick answers that people have given me over the years: "We let all our parts soak for at least four-hours, six-hours, eight-hours," or the famous "overnight." The correct answer for your in-house measurement control system can be found in the age-old proverb: It's not the hours you put in but rather what you put in the hours.

Before an appropriate thermal stabilization process can be developed we must understand not just thermal expansion, but also thermal transfer. Heat present in an object is an effect; the cause is the thermal energy moving through the object. Thermal energy, like all forms of energy, has just one agenda: to move along the path of least resistance. When a part is brought in from a warm shop floor environment, or taken out of someone's pocket and placed in a controlled (we'll say 70° ± 2°F) measuring lab, the thermal transfer starts. How long it takes is determined from a few easy to understand (and monitor) variables.

As soon as a warm object is placed in contact with an object of a lower temperature (an object containing less thermal energy), the thermal transfer process begins (*Figure 7-4*). The thermal energy from the warmer object transfers into the cooler object until the thermal energy balances between the masses of the two objects. When this occurs, they are at a point of thermal stability with each other (they reach thermal equilibrium). The two objects have the same amount of thermal energy, which, once again, is the cause, and the effect is that both objects are now the same temperature all the way to their cores. This transfer process continues as the two objects transfer their thermal energy into the object they are in contact with: table, bench-top, etc., and eventually the energy is transferred into the atmosphere of the room. Think of the thermal energy as a 16-ounce bottle of pink lemonade that is added to a swimming pool. The lemonade is in the pool and doesn't "go away," it's just diluted so thin it can't be detected.

Different materials transfer thermal energy much like different objects transfer electricity. Materials such as steel, copper, and brass have great thermal conductivity, while materials such as wood, plastic, and granite often act more like insulators and resist the process. There are also variables within these variables that are often overlooked. For example, the thermal transfer is affected by the masses of the objects. If a warm object is bought in from a shop-floor environment and set on a 1" thick steel plate that measures 6" wide × 12" long, the time needed to reach thermal stability

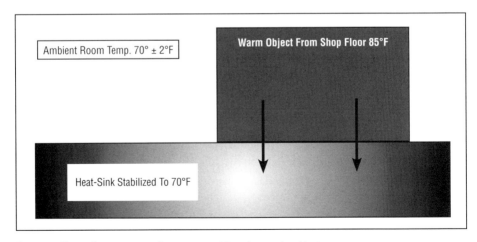

Figure 7-4: Thermal energy moves from warmer objects into cooler objects.

between the warm part and the steel plate (thermally stable to the inspection lab temperature) might be X minutes. If the same part were brought into the inspection lab and set on a 1" steel plate that measured 12" wide × 12" long, the time needed to reach thermal stability might be closer to $X\div2$ minutes. **The thermal transfer process is much like a tug-of-war between two different objects that have rates of thermal transfer that are proportionate to the mass of each object.**

With this in mind, let's look at the original question: "How long should we let our parts soak?" and discuss a few examples. Let's imagine for a moment that I walk in from the shop floor in August and remove two nearly identical sized plug gages from my pocket, where they have been all morning. Once inside the inspection lab (70° ± 2°F), I place one of the plug gages on a heavy duty steel file cabinet and the other on a gage repair work bench that has a very thick, butcher block type wood tabletop surface.

Now… how long will each object need to stabilize to the ambient temperature of the inspection lab? It becomes a question that can't be answered because I have not defined the variables:

1. How large are the plug gages (approximate mass)?

2. How warm are the plug gages (approximant temperature or amount of thermal energy)?

3. Are the plug gages in contact with a thermal conductor (steel table) or thermal insulator (wooden workbench)?

4. How large (mass) is the surface each plug gage is in contact with?

Even the famous trump card of "overnight" still cannot be proven as an adequate amount of time for the plug gages to stabilize.

The best example of how thermal energy travels though different material, based on the material's mass, can be found in your kitchen. If you're baking a cake and you open the door of a 400°F oven you'll feel the heat from the oven. You can reach into the oven and wave your hand around in the 400°F air without much discomfort. This is possible because the air has very little density, so it has very little mass, and there is a low level of thermal energy needed to maintain the temperature within the air. If you touch the surface of the baked cake you will notice the sensation of the temperature more, because the cake has a higher density than the air. This gives it more mass, which means there is more thermal energy moving through the cake to maintain its temperature of 400°F. If you accidentally touch the edge of the cake pan, you'll really notice its 400°F temperature because the metal cake pan has a much higher level of density than the cake or the air. This gives it more mass, which means there is much more thermal energy moving through the metal to maintain its temperature. Each material (the air, the cake, and the metal pan) balance out by only transferring enough thermal energy to attain thermal stability with each other.

Now, the answer of how long will an object need to stabilize to the ambient temperature of your inspection lab lies in a simple test that only requires a calibrated thermometer that can be purchased for well under $200. This thermometer can be used to control the temperature of most, if not all, of your future measurements. Keep this in mind:

"The supplier shall… ensure the environmental conditions are suitable for the calibrations, inspections, measurements, and tests being carried out."

In fact, they can be met and exceeded with a $200 thermometer and a $5 calculator. It's a simple science-fair type project that can save thousands of dollars in wasted time later. Here's how it works:

In your inspection laboratory setup, you most likely use some object that parts are placed on to reach thermal stability. This is commonly called a "heat sink." I would recommend trying a 1" thick piece of steel, approximately 12" × 18", surface ground smooth on both faces. Attached to the heat sink should be your thermometer. This ensures that you are monitoring the temperature of the heat sink. Many places think the way to go is along the lines of an environmental recorder placed near the center of the lab. I don't recommend this because controlling the ambient temperature is very difficult as temperature shifts of 5°F are very common during a period of a couple of hours. Furthermore, part core-temperature shifts are much less drastic within the room because the thermal transfer process between the ambient air and the steel part is very slow. The heat sink temperature can be monitored by one of two methods. A digital thermometer with a copper wire probe can be attached, using either adhesive tape or putty. The other option is quite a bit more basic, and that's the reason I prefer it. I recommend placing a glass beaker of mineral oil on the corner of your heat sink. It doesn't need to be laboratory grade mineral oil; a bottle from the dollar store will work just fine. A liquid in glass thermometer can be kept immersed in the mineral oil to monitor the temperature of the oil, which will remain thermally stable to the heat sink if not disturbed. This is seen in *Figure 7-5*.

Why do I recommend the most basic (almost archaic) of the two methods? Be-

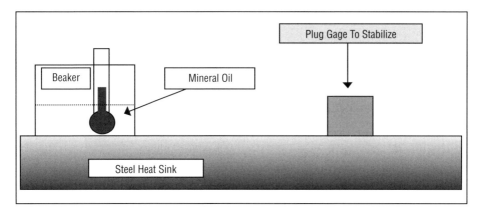

Figure 7-5: Monitoring the temperature of a "heat sink" using a liquid in glass thermometer.

cause it would have worked just fine for Isaac Newton, that's why. Think about it, the Egyptians designed and built the most difficult engineering wonders in the history of the world. Did they have CAD software, wireless communication, or global positioning satellites, of course not. They knew how to do it by counting beads on a string. If I was a potential customer walking into your inspection lab and I saw a $1000 digital thermometer attached to your heat sink, I'd think you spent some money. If I saw a $100 thermometer sticking out from a beaker of fluid <u>and you could explain to me how it all worked</u>, I'd be impressed, and I'd remember it for a long time. Now, back to the science-fair.

Place a plug gage (approximately Ø1.0000") on the heat sink and, for about a week, document the temperature of the heat sink (0.1°F increments if possible) once every 30-minutes. This will accomplish two things; first, it gives you a baseline to determine just how well you are controlling the <u>temperature of the objects</u> measured in your lab, and, second, you're quite certain the 1" plug gage has reached thermal stability with the heat sink.

Using the finest resolution instrument available, measure the diameter of the plug gage across a fixed point several times, and zero-set the same instrument to this fixed point. The next step is going to sound crazy, but place the plug gage inside a coffee pot and run water through the coffee pot (without coffee) with the plug gage set inside the pot (on the burner) for about 30-minutes. This purpose of this is simple: the hot water in the pot is over 150°F; by leaving the plug gage submersed we are very certain the gage core temperature will be <u>well above</u> anything we'll experience on the shop floor.

After 30-minutes, remove the plug gage, dry it quickly, measure it across the fixed point, and document your reading. You'll see the thermal expansion in the expanded reading. Place the plug gage in contact with the heat sink and then, after waiting one-minute, measure it again: it will be smaller. Continue measuring the plug gage at one-minute increments until it "shrinks" back to its original size, which is zero on your instrument. This thermal transfer process will most likely take less than 30-minutes.

Now you've established your own coefficient of thermal stability for your in-house measurement control. Because we heated the plug gage up to somewhat of an unrealistic temperature, we have also built a "fudge-factor" into our equation. We have proven that with the variables present in our situation, a 1" plug gage brought into our lab and placed on our 1" × 12" × 18" heat sink will reach thermal stability in X minutes.

I'm not going to profess to hold any type of a degree in physics or, more specifically, thermodynamics, so I suggest you perform similar tests on larger objects to confirm the calculated correlation and, when in doubt, always estimate on the side of a longer stabilization time. Also, when the stabilization test is performed, I would recommend using an instrument having a resolution of submicrons, such as 0.0005mm (or .00002"). If your facility does not have an instrument of this type resolution in your inventory, I suggest you arrange a three- to five-day loan, or a gage demo. You'll find this to be a very useful tool. I try and demo at least three or four new gage concepts per year just to have some Aces up my sleeve when a new project is in the works. Most reputable instrument companies are very eager to perform this service, because, if nothing else, they have another individual in the global market that can give first hand testimony about their products.

Thermal effects can be a hot topic, but if you know the math and science of the game, you need not get burned.

Chapter 8
What You Can't (Not) Know About SPC

Figures can lie and liars can figure…

Statistical Process Control (SPC) has been used for quite some time in industrial environments. There is a perception that it is always a manufacturing process. We have a bad habit of fitting concepts into a box and never thinking the concept can exist outside the box. For example, some of the same statistical tools used in SPC are also used in Measurement Systems Analysis (MSA), or "Gage R&R," yet few facilities view them as parallel. In fact, some facilities believe that SPC fits under the umbrella of a Production Control (however termed) department, while MSA is placed within the scope of the Quality Department. If this structure is used, the right hand not only doesn't know what the left hand is doing, it doesn't even know the left hand is there. A good example of this argument is the Automotive Industry Action Group (AIAG), which fails to reference SPC anywhere in Element 4.11

Almost everything we do repeatedly in life involves a process. Measurement in an industrial facility is, in and of itself, a process and a very delicate one at that. In order to successfully implement a MSA program, it helps to have fundamental knowledge of some simple SPC concepts. This can provide you with more than just the ability to satisfy a standard. When simple SPC concepts are implemented and a dilemma occurs at your facility, the measurement instrument—if not the root cause—can be eliminated as the problem almost immediately.

The first tool we should become familiar with is the **histogram** (*Figure 8-1*). A histogram is a graph that provides a visual image of the spread, range, variation, or whatever term you prefer, within a sample. Our samples will always contain data from a group of measurements. The histogram demonstrates **normal distribution**, which is assumed with measurement data. Normal distribution is sometimes called a "Bell Curve" because of the shape of the curve that is usually formed; most of the sample readings fall in an area near the mean (or average), and as we move farther away from the mean the number of readings decreases. The spread of the data can be divided into six sections called **sigma** or **standard deviations**. In a normal distribution, 68% of the readings will fall within one-sigma of the mean, 95% of the readings will fall within two-sigma of the mean, and 99.97% of the readings will fall within three-sigma of the mean. It may appear that 100% of the readings fall within three-sigma of the mean, but they never will because of one simple fact: a sample will never contain as much variation as the population. This is extremely true with histograms, because the population could become an infinite number of measurements in a certain application.

A good example of how a histogram could be used to control or confirm a measurement process would be a Rockwell Hardness tester in an industrial inspection laboratory that is used to confirm the heat-treating of our products or the products from customers. If you wanted to be able to confirm readings at any time between the annual calibration of the hardness tester, you could develop a histogram using a hardness test block. This sample will be used as a **reference standard** after the statistical tests are performed. If you're not familiar with hardness testing, continue to follow along and this example will still make sense. You could measure the hardness test block (using the C scale) three times in a row for 15 days. After the 15 days, you would have 45 total measurements and 15 subgroups of three readings.

If you constructed a histogram, you could determine the sigma or standard deviation and then determine the **probability** of how stable your readings were and how much variation is within the measurement process. For example; if you measured your test part three times each day for 15 days and you attained the following data, you could find out many things about your tester.

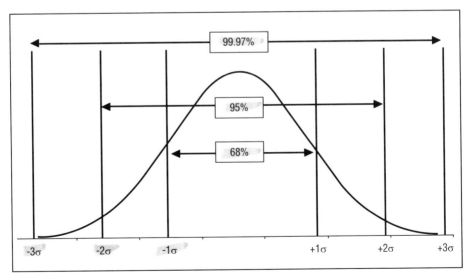

Figure 8-1: The Histogram. Image courtesy of SPC for Excel by BPI Consulting.

Day	1#	2#	3#
1	23.10	23.40	23.90
2	23.90	23.30	23.60
3	23.70	24.10	24.00
4	23.70	24.10	24.00
5	24.10	23.70	24.00
6	23.60	23.70	24.10
7	24.00	24.10	24.50
8	23.50	23.70	24.10
9	23.10	23.90	24.30
10	23.60	24.10	24.00
11	23.50	23.10	23.70
12	23.50	23.90	24.20
13	23.10	23.90	24.20
14	23.10	22.90	24.10
15	24.10	24.20	23.90

Your 45-reading average (or mean) attained was 23.784. The one-standard deviation (or sigma value) was 0.38. Following the rules of normal distribution, you could determine the following:

- 68% of the time the measured value fell within one-sigma of our mean (23.40 – 24.17)

- 95% of the time the measured value fell within two-sigma of the mean (23.02 – 24.55)

- 99.97% of the time the measured value will fall within three-sigma of the mean (22.64 – 24.93).

If you want to add a visual effect, you can display the histogram (*Figure 8-2*) near the hardness tester.

This histogram is such a valuable tool because it gives inspection personnel a very fast and accurate method to confirm an instrument when a less than favorable reading is attained on a production part. If the measurement is questioned, the inspector can measure the reference standard and compare its measured value today to the measured value during the study. I would suggest performing these studies after an outside laboratory calibration. This will provide a good baseline to use throughout the year. If any time during the year you notice an unac-

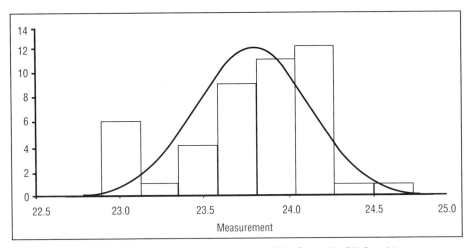

Figure 8-2: Histogram of hardness tester data. Image courtesy of SPC for Excel by BPI Consulting.

ceptable "drift" in measurement results, you should be able to detect it very quickly.

You could use the same technique to keep an eye on a CMM that is used frequently for cylinder or bore measurements on milled part features. If you measured a 2" Ring gage three times per day (using the same techniques, number of points, etc.) for 15 days after a calibration was performed, you could determine a baseline for the CMM's average and standard deviation from the sample data.

Day	1#	2#	3#
1	1.99999	1.99999	2.00003
2	1.99997	1.99989	1.99997
3	1.99993	2.00003	2.00003
4	1.99983	1.99988	1.99987
5	1.99978	1.99980	1.99978
6	2.00000	2.00004	2.00006
7	1.99992	2.00019	2.00019
8	1.99998	1.99997	2.00006
9	1.99985	1.99997	1.99983
10	1.99996	1.99992	1.99997
11	2.00021	2.00026	2.00020
12	1.99980	2.00003	2.00003
13	2.00011	2.00013	2.00012
14	2.00019	2.00017	2.00014
15	1.99988	1.99994	2.00002

The 45 reading average or mean attained 2.00000", which is not surprising. The one-standard deviation or sigma value was .00013" (a tenth and 30 millionths). Following the rules of normal distribution, you could determine the following:

- 68% of the time the measured value fell within one-sigma (1.99987" – 2.00012")
- 95% of the time the measured value fell within two-sigma (1.99974" – 2.00025")
- 99.97% of the time the measured value will fall within three-sigma (1.99962" – 2.00038").

This data gives us some interesting information: The overall average was as close as it could get, but the variation within the sample data is probably not as good as most people (who do not have much CMM experience) would think. This becomes very clear with the help of the chart shown in *Figure 8-3*. Now, when an operator measures the 2" ring gage reference standard, he or she can take one look at their result and one look at the chart and decide how today's probe calibration "measures up" to what the machine's normal average.

The histogram also works very well to document the 1" master sphere measurement taken by a CMM after a probe or stylus has been calibrated to the masterball. Normally, a CMM will almost have its own "fingerprint" if you measure a sphere (taken from no fewer than nine data points) after the probe has been calibrated. Some CMMs will average .99990" after hundreds of measurements taken over the span of a year. This information becomes very helpful, because in measurement science it becomes beneficial to view SPC data as a metrologist, not as a statistician. This is a concept that needs to be demonstrated in order to be fully explained.

Let's assume for a moment that we have performed 75 sample measurements on a 1" masterball over a time span of several months. The SPC study, as shown in *Figure 8-4*, has yielded the following data.

- Our 75 reading average or mean attained 999878"
- 68% of the time or one-sigma (between .999790" – .999966")
- 95% of the time or two-sigma (between .999702" – 1.000054")
- 99.97% of the time or three-sigma (between .999614" – 1.000142").

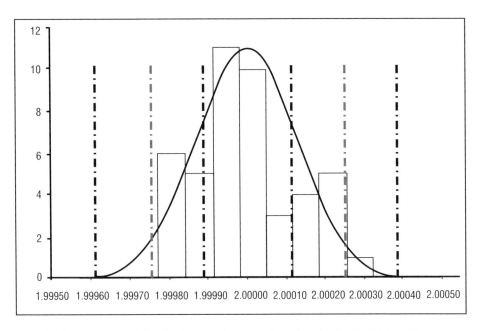

Figure 8-3: Histogram of CMM data (2" ring gage). Image courtesy of SPC for Excel by BPI Consulting.

Now, if we calibrated a stylus and confirmed the calibration by performing a sphere measurement (nine-data points) of our 1" masterball to obtain a reading of 1.00005", a statistician would tell us to go with it because our process was "in-control." As a metrologist, I would have to argue this may not be the best move. From an SPC perspective, the process is in control, no doubt about that. But the real world process (in this situation) is not probe calibration, it's using the CMM to confirm our parts. A metrologist would be a fool to measure parts (with his or her reputation on the line) without getting the CMM "dialed in" as best as they could. Does this mean I would spend an hour recalibrating a probe until the sphere measurement was exactly .999878" (our 75 reading average), of course not? I would, without a doubt, spend five to ten more minutes to try and tweak the probe calibration closer to within one-sigma of our average.

Keep in mind that this is another practice that will have its share of critics, such as the machinist standing over the inspector's shoulder looking at his wristwatch. Just remember to never let someone else's impatience affect your judgment. If you do, and the customer rejects the part, the machinist will never hold his hand up and say "my fault boss, I didn't allow him to get the CMM set-up correctly." What a

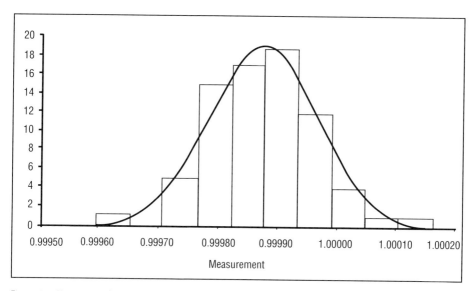

Figure 8-4: Histogram of CMM data (1" master sphere). Image courtesy of SPC for Excel by BPI Consulting.

great world we'd work in if inspectors were allowed to stand behind a machinist working at his or her mill and give them advice about how long they were taking. No chance at that... machine trades are an art and measuring is just 1-2-3-enter... yeah, right!

Another chart that can be very beneficial is the X-bar and R chart. **X-bar** is another term for average and **range** is the variation within the subgroups; this chart can also detect range between the subgroups. The X-bar and R chart is commonly used to record measurements from a production process to determine if the process variation is within control limits (in control) or outside of control limits (out of control). When a production process is being analyzed using X-bar and R charts, the goal is to keep the process in control.

You can use an X-bar and R chart to analyze a Measurement System in much the same manner. By using the measurements of an artifact, such as the 1" masterball data used to develop a histogram, the X-bar and R chart can determine how consistent this measurement is over the entire time span used to collect the data. Since the same object was being measured, a good measurement system or process will display a consistent average with very little range variation. This picture truly does paint a thousand words.

The X-bar and R chart displays many items about the measurement, as shown in

Figure 8-5. The three-reading averages almost all fall within .0001" of the overall average of .99990" and the average range is .0001". But, for some reason, we are definitely experiencing a downward trend within the daily three-reading average. We'll want to keep an eye on this and investigate further if it continues.

Another concept we should discuss involves using random sample parts which vary somewhat in their dimensional values. When these samples are used, it requires the X-bar and R chart to be interpreted backwards. The reason for this is we want a Measurement System to be able to determine the variation (or range) between the part values. There is a very common misperception when dealing with measurement variation within sample parts: people tend to think that when instrument readings depict variation within measured results, there must be something "wrong" with the instrument. In reality, the instrument is detecting the variation that is present, just as it should. If we use random parts (measured at a fixed point), **we should expect an effective measurement process to determine the smaller parts from the larger parts from the "middle of the road" parts.** To demonstrate this concept, we can look at sample data derived from a study in which 15 sample parts were measured three-times each (*Figure 8-6*). We'll discuss blind studies more in the next chapter, but for now just

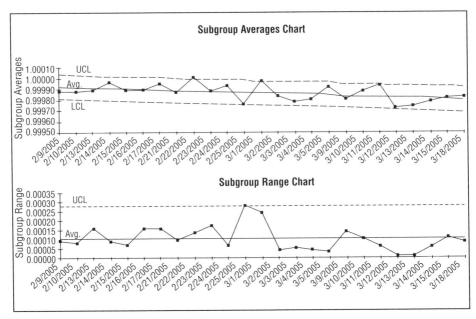

Figure 8-5: X-bar and R chart of CMM data (1" master sphere). Image courtesy of SPC for Excel by BPI Consulting.

remember our parts were not labeled in a way that would allow our measurer to determine which parts were which, but rather they just performed 45 measurements.

If we chart the data using an X-bar and R chart, we can see the variation between the sample parts (*Figure 8-7*). This is where we need to analyze the chart much differently. A pure statistician might look at the average chart and determine the "process" is not in control, but the "process" we are analyzing is a <u>measurement</u> process and we want it to be capable of sensing the deviation between the parts of different sizes. We can look at the range chart and determine the measurement system is very stable because the average range within the subgroups is just over a tenth. This is very good at a measurement (or instrument) range of 4.250".

If you work with these two charts (histograms and X-bar/R charts) and become familiar with how the numbers are crunched within the charts, your ability to implement and understand the techniques discussed in the next three chapters will be accomplished within a much shorter learning curve. With some simple planning, a data sheet could be created and the sample measurements could be taken on several measurement systems in only a few minutes a day. Before you know it, you'll have three- or four-weeks of very solid data to help you understand exactly "what you've got" in a given measurement. Some will insist they don't have the time to do it, but the winners are those who realize they don't have time not to do it.

#1	#2	#3
4.2511	4.2510	4.2511
4.2508	4.2508	4.2507
4.2503	4.2503	4.2502
4.2505	4.2506	4.2507
4.2514	4.2512	4.2512
4.2510	4.2509	4.2509
4.2512	4.2511	4.2511
4.2502	4.2503	4.2503
4.2508	4.2507	4.2508
4.2511	4.2511	4.2511
4.2502	4.2502	4.2501
4.2509	4.2508	4.2508
4.2507	4.2505	4.2505
4.2511	4.2512	4.2513
4.2509	4.2508	4.2509

Figure 8-6: Data from 45 "blind" measurements.

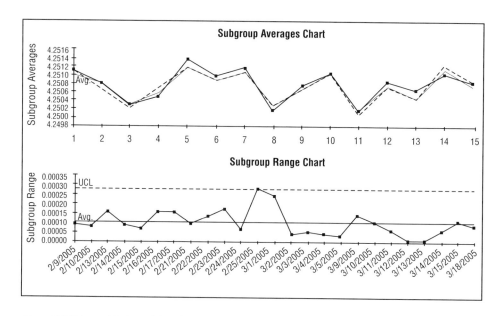

Figure 8-7: X-bar and R chart of data derived from 45 "blind" measurements. Image courtesy of SPC for Excel by BPI Consulting.

Chapter 9
Introduction to Measurement Systems Analysis (Gage Repeatability and Reproducibility)

No two things are exactly alike… and even if they were, we'd obtain different results when we measured them…

QS-9000 and TS-16949 are governed by AIAG (Automotive Industry Action Group) standards that require:

"…Statistical studies shall be conducted to analyze the variation present in the results of each type of measurement and test equipment. These methods and acceptance criteria used should conform to those listed in the Measurement Systems Analysis (MSA) reference manual version 3."

An in-depth tutorial on MSAs could easily become a book in itself. In this chapter we will develop an overview of what MSAs are evaluating within a measurement. What repeatability and reproducibility are, and how they interact and affect each other will be discussed, as well as how common MSA studies are set up, performed, and evaluated.

Repeatability is defined as the range (or variation) of readings obtained when one measurement instrument is used several times by one operator when measuring the identical characteristic of the same part (*Figure 9-1*). This is often called EV for "equipment variation."

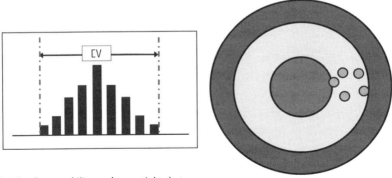

Figure 9-1: Good repeatability produces a tight shot-group.

Reproducibility is defined as the range (or variation) between the averages of measurements made by separate operators when one measurement instrument is used several times by separate operators to measure the identical characteristic of the same part (*Figure 9-2*). This is often called the within appraiser variation or AV for "appraiser variation."

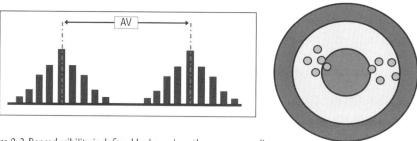

Figure 9-2: Reproducibility is defined by how close the average readings are.

It's worth noting some of the misperceptions about measurement. What is sometimes termed as **accuracy**, or inaccuracy, is actually **precision**, or lack of precision. If you asked the average person if the shooter in *Figure 9-3* was accurate, they would probably say no, because the shooter is all over the target. However, if we averaged all of the shots, the shooter would average very near the bull's-eye.

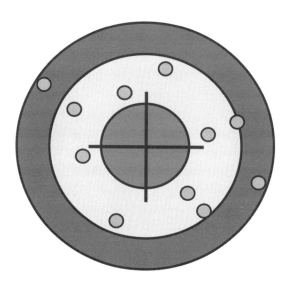

Figure 9-3: Accurate, but not precise.

What most think of as accuracy is actually precision, or the ability to place a shot over and over in the same place, as seen in *Figure 9-4*. When a shooter gets really "dialed-in" they are both accurate and precise, as seen in *Figure 9-5*.

When a measurement is not accurate, it has bias, which is the technical term for error. Some MSA studies are used to estimate how much bias (error) is within a measurement, and some MSA studies are used to determine how much variation is

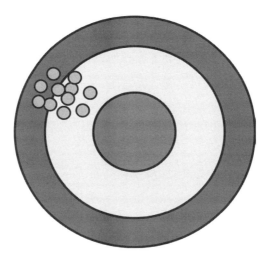

Figure 9-4: Precise, but not accurate.

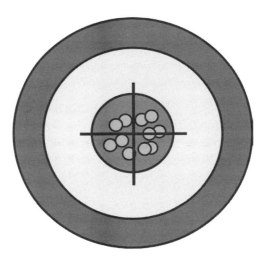

Figure 9-5: Accurate and precise.

contained within the measurement results. The variation from both repeatability and reproducibility are quantified and compared to the specification of the characteristic being measured to determine how much of the specification is consumed by measurement variation.

The MSA (version 3) recommends that, to be considered acceptable, less than 10% of the specification be consumed by variation within the measurement system. If between 10% and 30% of the specification is consumed by measurement system variation, it may be acceptable but should be improved. When more than 30% of the specification is consumed by variation, the measurement system is considered unacceptable and must be improved. This is a very strict requirement that we will discuss in greater detail later, but for now we'll make sure we are clear by discussing some examples.

- If our MSA variation is: 0.015mm

    And our part tolerance is: 0.05mm

    Our MSA variation consumes 30% of our part specification.

- If our MSA variation is: 0.015mm

    And our part tolerance is: 0.1mm

    Our MSA variation consumes 15% of our part specification.

- If our MSA variation is: .0012"

    And our part tolerance is: .005"

    Our MSA variation consumes 24% of our part specification.

- If our MSA variation is: .0012"

    And our part tolerance is: .010"

    Our MSA variation consumes 12% of our part specification.

The correlation between part specification and MSA percentage is a very easy concept to grasp, but you'll be amazed how misunderstood it can become within your chain-of-command. Part of this misunderstanding occurs because it is human nature to think that Measurement = Gage. When an MSA using a specific gage type (such as a bore gage along with dial indicator and ring gage master) produces an MSA result of 0.015mm (15% of a 0.1mm tolerance), people tend to keep those two items in their head :

Bore gage = 15% MSA.

Later on, when taking another measurement using a bore gage with a completely different measurement and tolerance, a MSA result of 0.021mm (42% of a 0.05mm tolerance) is achieved, and they think something went wrong. Nothing went wrong, everything in the measurement changed except the instrument. We'll discuss these types of scenarios in the next chapter in more depth, but it makes sense to touch on them now.

We should discuss terminology before going any further. Different entities (or references) sometimes have different terminology. In the MSA reference manual, any variation or uncertainty found within the measurement results is termed **gage error** or **measurement system error**. This makes perfect sense when we view the results from a perspective that whether the uncertainty is bias or variation, it is causing errors in our ability to consistently measure our parts.

Common MSA study formats usually include **linearity studies** which document an instrument's bias (or accuracy) throughout a specific measurement range. **Stability studies** document a measurement system's bias over a given timeframe. **Average and range studies** document the amount of variation (both EV and AV) within a measurement system. A short **range study** is a very quick study used to determine either: approximately how much repeatability variation is within a given system, or if the variation has changed since a past MSA study was performed.

Range studies are very valuable tools that will save a lot of time during test and retest, but they should never be considered as a complete MSA study. As you'll learn later in this chapter, range studies are quick "snapshots" as opposed to a thorough analysis. We will discuss range studies first because they are a good starting point.

When we look at the histogram examples of the definitions of repeatability and reproducibility, we should always accept the philosophy that a measurement system must be able to repeat before we can determine if it will reproduce. If the precision of the measurement system is bad enough, or the precision of the "shots" are so scattered, the averages will most likely agree. If you look at the target in *Figure 9-6*, you'll see that both shooters (white and dark-gray paintballs) are all over the target. This means they can't repeat their shots. But, if we average their shot groups, we would get a false report that they could reproduce. After we get each shooter to tighten up their shot groups (attain repeatability) then we can (truly) see how much reproducibility we do or do not have.

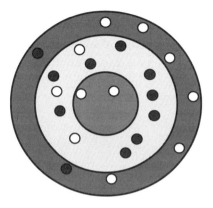
Averages between white and black are very close, so they would appear to reproduce, yet neither shooter is any good.

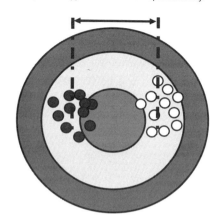
As their shot groups tighten up (better repeatability) we see the real reproducibility.

Figure 9-6: Poor repeatability can produce false reproducibility.

This is where a short range study can save so much time. Average and range studies are very time-consuming to perform. It's a much wiser strategic move to perform a range study, which will tell you very quickly if you have a repeatability problem, before you waste the time and effort needed to carry out an in-depth average and range study. These studies are just ballpark estimates to save time, so don't be afraid to experiment with slightly different test sample sizes as you gain experience. For our example range study we'll use two sample parts and two operators. Sample parts are almost always "sample parts"—that is, not special parts, not production master parts, not parts we know are in the middle of our specification, they're just **random, sample parts**. All MSA measuring should be *blind*, not as in with your eyes closed, but as in "the operators don't know the difference between the sample parts." The operators only need to know that you handed them a part and they measured it, and then you handed them another part and they measured it. Operator #A shouldn't know ahead of time what value Operator #B obtained on any of the measurements. If we have any preconceived notions about what the parts will (or should) measure, our test results should be considered skewed and invalid.

We'll have each operator measure Parts #1 and #2 five times each, alternating each time. Part #1 will also be used as Part #3, #5, #7, #9 and Part #2 will also serve as Part #4, #6, #8, #10. Again, each operator should be given these parts discreetly so it appears they are measuring 10 different parts. After the measurements are documented, we have 10 readings from two operators for Part #1, and 10 readings from two operators for Part #2. We'll use data taken from Part #1 measurements for our study. If we want to double check our study results, we can use the data from Part #2. (The formula calculations from a short range study can be seen in *Figure 9-7*.)

Formulas

Short Method

$$R \& R = 5.15 \times d_2 \times \overline{R}$$

Operators = 2, Parts = 5, d_2 = 1.19

Operators = 2, Parts = 10, d_2 = 1.16

Operators = 3, Parts = 5, d_2 = 1.74

Operators = 2, Parts = 10, d_2 = 1.72

$$\%R \& R = 100 \times R \& R \div Tol$$

Figure 9-7: **Formula calculations for short range study.** Image courtesy Easy Gage R & R by Math Options, Inc.

Let's look at our data for Part #1.

Operator A	Operator B
25.22mm	25.20mm
25.22mm	25.22mm
25.22mm	25.21mm
25.21mm	25.20mm
25.20mm	25.22mm

This is where the part specification plays an important role. If the specification of this characteristic were 25.3mm ± 0.3mm, our range (or gage error in this study) of 0.052mm would consume 8.66% of our part tolerance, as shown in *Figure 9-8*. That's great.

If, on the other hand, our part specification were 25.3 ± 0.15mm, our range would consume 17.31% of our tolerance (*Figure 9-9*).

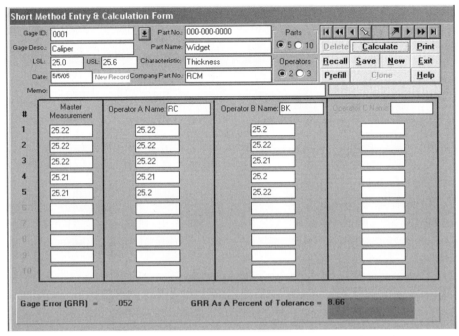

Figure 9-8: Example of a short range study. Image courtesy Easy Gage R & R by Math Options, Inc.

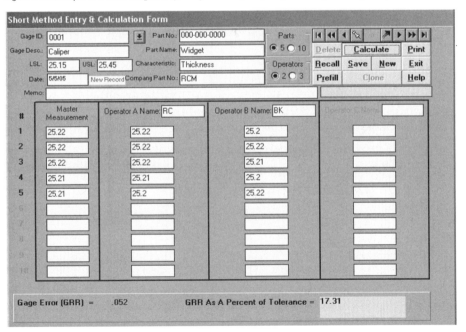

Figure 9-9: Repeatability consumes 17.31% of part tolerance. Image courtesy Easy Gage R & R by Math Options, Inc.

Finally, if our part tolerance were 25.3 ± 0.01mm, our range would consume 25.97% of our part tolerance. Wow!

This is worth bringing up another strategic concept. If you look at the internal calibration accuracy tolerances we discussed in Chapter 3, you could determine how much error (or accuracy) we could be "giving up" before the instrument leaves the calibration lab. You could develop a matrix to list the different shop floor instruments and the percentage of tolerances that are most likely "sacrificed" before a single part is measured in an MSA study. Take, for example, a run-of-the-mill dial caliper with a manufacturer's accuracy tolerance of ± .001" (or 0.02mm). Our range of readings at any given dimension during calibration could vary .002" (or 0.04mm). You can plug this range into different part tolerance levels to give an estimate of what's the best that could be expected of this instrument on the shop floor:

Part tolerance .020" = 10%

Part tolerance .010" = 20%

Part tolerance .005" = 40%

By contrast, if you look at the accuracy of a 1" micrometer (±.0001") it compares much better:

Part tolerance .020" = 1%

Part tolerance .010" = 2%

Part tolerance .005" = 4%

Part tolerance .004" = 5%

Part tolerance .002" = 10%

Keep in mind these percentages represent the <u>possible errors</u> obtained during calibration, without having measured a part on the shop floor. No doubt about it, the (less than) 10% spelled out in the MSA reference manual is World Class!

Linearity studies document an instrument's bias (or accuracy) throughout a specific measurement range. If you think about it, a well thought out calibration procedure should test the instrument throughout its full measurement range. For example, a 1" micrometer can be calibrated by measuring these five gage blocks: .100"; .310"; .515"; .720"; .990". Or a 0-25mm micrometer can be calibrated by measuring these gage blocks: 5.10mm; 10.20mm; 15.30mm; 20.40mm; 25.00mm. This is a very pure form of a linearity study, but some might feel it doesn't exactly fit into the MSA category because we must use sample parts (as I stressed earlier), not calibration artifacts.

Most MSA studies are best understood by walking through a sample test. To demonstrate, we'll use a minimum of five samples that (due to process variation) encompass the entire operating range of the instrument. Before the test is conducted, we'll establish accepted reference values for each sample by layout inspection. The test is very simple: Have one operator measure each sample 10 times (blind, in random order) and document the results. Using the 10 reading averages of each sample part (obtained during the test), calculate the bias of the instrument at the range of each reference value. The importance of using a blind study cannot be overstressed. If the parts were labeled and a measurer realizes that "Part #3" measured 10.21mm twice in a row, then the part will—almost certainly—obtain a value of 10.21mm each of the remaining eight times the measurer is handed Part #3. If, by chance, a measurement of Part #3 obtained any other reading but 10.21mm, the operator would either tell you they got 10.21mm, or "adjust" the measurement to obtain 10.21mm. This would not make them dishonest, just human. The instant a measurer in an MSA study has a preconceived knowledge of what a sample part "should" measure, the test should be considered tainted.

Figure 9-10 displays the graphical result of a sample linearity MSA study. In this very quick study, a 0.001mm resolution digital micrometer was used to measure sample parts throughout its 25mm measuring range (any similarity to these part values

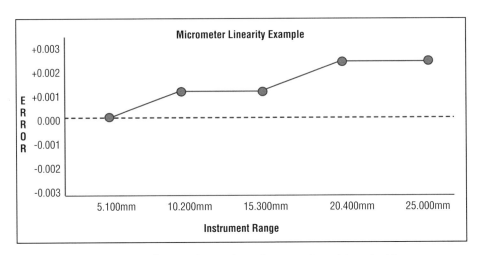

Figure 9-10: Linearity is the error (or increasing error) as an instrument is used throughout its range.

and the calibration points listed earlier is strictly coincidental). This study displays that as we measure further into the range of the instrument, the inaccuracy increases. Normally, in dimensional measurement this occurs more often than it does not.

For a more in-depth MSA, a facility can use an **average and range study**. This usually involves more sample parts, more sample measurements, includes two to three operators performing measurements, and is sometime referred to as an AIAG Long Study. Within the "Big Three," the calculations involved within the MSA vary slightly.

AIAG Long Method

Equipment Variation (E.V.) - Repeatability

E.V. = $\bar{R} \times K1$

   Trials = 2, K1 = 4.56
   Trials = 3, K1 = 3.05

%E.V. = 100 × (E.V. ÷ T.V.)

Appraiser Variation (A.V.) - Repeatability

A.V. = $\sqrt{(\bar{X}\text{diff} \times K2)^2 - ((E.V.)^2 \div (n \times r))}$

   n = number of parts, r = number of trials
   Operators = 2, K2 = 3.65   Operators = 3, K2 = 2.70

   Note: If the calculated A.V. is less than zero, the A.V. is reported as zero.

%A.V. = 100 × (A.V. ÷ T.V.)

Repeatability & Reproducibility (R & R)

R & R = $\sqrt{(E.V.)^2 + (A.V.)^2}$

% R & R = 100 × (T.V. ÷ Tolerance)

Part Variation (P.V.)

P.V. = Rp × K3

   Parts = 5, K3 = 208   Parts = 10, K3 = 1.62
   Rp = $\bar{X}$p max - $\bar{X}$p min
   $\bar{X}$p values are the *averages* of each measured part

Total Variation (T.V.)

T.V. = $\sqrt{(R \& R)^2 + (A.V.)^2}$

Figure 9-11: Formula calculations for (AIAG) average and range (long) study. Image courtesy Easy Gage R & R by Math Options, Inc.

GM/Chrysler Long Method

Equipment Variation (E.V.) - Repeatability

E.V. = $\bar{R} \times K1$

  Trials = 2, K1 = 4.56   Trials = 3, K1 = 3.05

%E.V. = 100 × (E.V. ÷ T.V.)

Appraiser Variation (A.V.) - Repeatability

A.V. = $\sqrt{(\overline{X}diff \times K2)^2 - ((E.V.)^2 \div (n \times r))}$

  n = number of parts, r = number of trials

  Operators = 2, K2 = 3.65   Operators = 3, K2 = 2.70

Note: If the calculated A.V. is less than zero, the A.V. is reported as zero.

%A.V. = 100 × (A.V. ÷ T.V.)

Repeatability & Reproducibility (R & R)

R & R = $\sqrt{(E.V.)^2 + (A.V.)^2}$

% R & R = 100 × (R & R ÷ Tolerance)

Figure 9-12: Formula calculations for (GM and Chrysler) average and range (long) study. Image courtesy Easy Gage R & R by Math Options, Inc.

Ford Long Method

Equipment Variation (E.V.) - Repeatability

E.V. = $\bar{R} \times K1$

  Trials = 2, K1 = 4.56   Trials = 3, K1 = 3.05

%E.V. = 100 × [(E.V.)² ÷ (R & R × Tol)]

Appraiser Variation (A.V.) - Repeatability

A.V. = $\overline{X}diff \times K2$

  n = number of parts, r = number of trials

  Operators = 2, K2 = 3.65   Operators = 3, K2 = 2.70

Note: If the calculated A.V. is less than zero, the A.V. is reported as zero.

%A.V. = 100 × [(A.V.)² ÷ (R & R × Tol)]

Repeatability & Reproducibility (R & R)

R & R = $\sqrt{(E.V.)^2 + (A.V.)^2}$

% R & R = %E.V. + %A.V.

Figure 9-13: Formula calculations for (Ford) average and range (long) study. Image courtesy Easy Gage R & R by Math Options, Inc.

These differences can produce an extremely subtle effect on the overall outcome of the MSA study: you can see the formulas in *Figures 9-11, 9-12,* and *9-13*.

The "AIAG Method Data Entry & Calculation Form," detailing test results, is shown in *Figure 9-14*. This MSA format tells us a lot about our measurement. For starters, we have an insignificant amount of reproducibility variation within the readings. This could mean one of two things:

1. The operators average the same because all three measure very well, and the measurement system is "dialed in" for all three individuals. If this is the case, we should see very low levels of repeatability variation also (*Figure 9-15*, left).

2. The operators average the same because their "shot groups" are spread out. If this is the case, we will see very high levels of repeatability variation (*Figure 9-15*, right).

In our sample average and range study we only have 0.04mm of repeatability variation within our results. This is between three operators and 90 total measure-

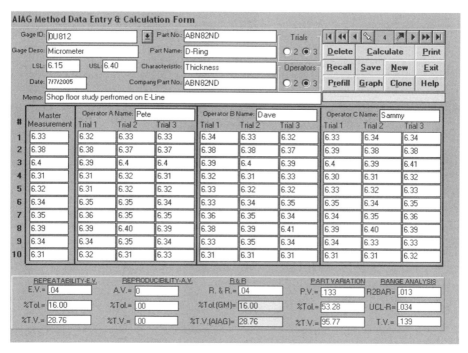

Figure 9-14: Example Average and Range study consumes 16% of part tolerance. Image courtesy Easy Gage R & R by Math Options, Inc.

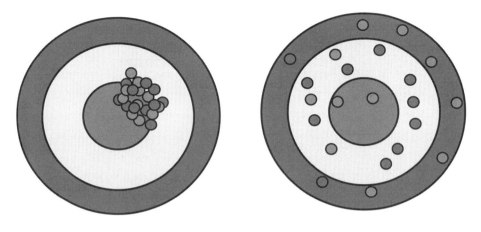

Figure 9-15: Both low levels of repeatability variation (left) and high levels of repeatability variation (left) can result in low levels of reproducibility within an MSA study result.

ments! This variation consumes 16% of our 0.25mm (6.15mm Min. / 6.40mm Max.) part tolerance.

These basic studies represent a good, solid start to an in-depth MSA program. If you can learn to plug measurement numbers into these three study types, you can begin to develop a very effective MSA system at your facility.

Only tomorrow knows what you'll discover.

Chapter 10
Elements of a Measurement

"So we fix our eyes not on what is seen but on what is unseen... for what is seen is temporary but what is unseen is eternal..."

Once an MSA program is developed and implemented, many measurements will not meet the strict requirements spelled out in QS-9000: 1998, and MSA version 3. These measurement systems will need to be analyzed, improved, and retested. In order to begin this process, we must come to grips with some common myths and misperceptions.

First of all, "Gage R&R" is a very misleading term. Measurement Systems Analysis involves testing the entire measurement system. When we walk out on the shop floor and see an operator measuring a part using an instrument (or gage), all we tend to see is the gage. What this usually means is if the measurements at a process are somewhat stable and it is easy to run a characteristic at a target value, then scrap and rework levels are low, and operators aren't disagreeing on parts that are at the tolerance limit (but good), or near the tolerance limit (but not good). In this situation, we say: "Great gage!"

On the other hand, if we have a process where we can't seem to control a characteristic, parts are being measured as "in tolerance" by some operators, only to be rejected later when measured by other operators, and scrap and rework are unacceptable, we say: "This gage is junk!" What we fail to realize is that our human nature is to look at an operator using an instrument to measure a part, and *only see* "the gage." In reality, there are six elements that are present in the majority of shop floor measurements.

This is where a "chess mentality" is the best approach. If you are in the quality or precision measurement field, and you don't play the game of chess, start today! There is no better tool to teach strategic thinking than learning the strategies involved in chess. The secret of mastering strategic thinking is being able to select the <u>best</u> move when faced with several possible moves. Every move must be thought out in a cause-effect scenario, and the best choice is usually an easy one. But you must think of every move as a domino effect. In the game of improving your MSA variation, we make moves to eliminate sources of variation. Instead of trying to capture pawns, knights, bishops, rooks, and queens, we'll try and lower the effects of variation from instruments, operators, methods, parts (or part geometry), environments, and tolerance sizes. Each element induces error, variation, and uncertainty within the measurement result(s), or, as with the size of tolerances, affects the magnitude of the variation that is present. We should review each element separately.

Instrument – Today's measuring instruments are expected to measure and repeat within increments of .0001", and sometimes .00005" (or 0.01mm and 0.001mm in Metric applications). Even in a near perfect environment, individual instruments can measure slightly different at these increments. As discussed in Chapter 3, all variable instruments have an acceptable amount of error that can be present (and expected). If a height gage has a mfg. accuracy specification of ±.001" (±0.02mm), it is permissible that height gage #1 can consistently measure a 4" gage block as 4.001" (or 100mm gage block as 100.02mm), and height gage #2 can consistently measure a 4" gage block as 3.999" (or 100mm gage block as 99.98mm). The gages themselves differ by .002" (or 0.04mm) and both are within accepted accuracy specifications. Also, some instruments can have combined uncertainties when measuring in certain applications. A caliper may have a mfg. accuracy specification of ±.001" (±0.02mm), which means caliper #1 can consistently measure a 4" gage block as 4.001" (or 100mm gage block as 100.02mm), and Caliper #2 can consistently measure a 4" gage block as 3.999" (or 100mm gage block as 99.98mm). But if we were measuring a Ø2" (or Ø50mm) plug gage we would be using the bottom portion of the caliper measuring jaws. Now we must take into account an additional .001" (or 0.02mm) possible parallelism of measuring face (jaws) error.

Operator(s) – Many measuring instruments or documented measurement methods go to great steps to take human interpretation out of a measurement result. No matter how much you try, you still have a person taking a measurement. No two people are alike. We all see things differently and feel things differently. We should always expect some level of operator-to-operator variation within our measurement results.

Methods – Different operators are sometimes comfortable with different methods. The textbook method for using a 1" (or 0–25mm) micrometer is to hold the instrument in one hand and the part in the other hand and take the measurement one-handed. Some operators are comfortable with this method but some may not seem coordinated enough and need to place the part on a stable surface and measure using both hands. These different methods may produce slightly different readings in some measurements. Now, the Devil's advocate might suggest everybody just use the same method. This might not be a wise chess move. If an operator uses a method they are not comfortable with, you might be adding more operator variation than you are eliminating method variation.

Parts (or part geometry) – Each characteristic of a customer's part may have its own level of difficulty when taking the needed measurement(s). Square or rectangular parts may seem much easier to measure than round parts. Also, in today's real-world resolution of microns (0.001mm) and millionths (.000050"), nothing is round, nothing is flat, and no two surfaces are parallel. Let's look at a very basic example. The best illustration is a part that (*Figure 10-1*) resembles a Ø1.000" (or 25.4mm) cylindrical shaft that is 6" (or 151.6mm) long. The drawing requires the diameter to be confirmed as well as the overall length. In the world of computer aided drawing (CAD), the shaft is always round (and cylindrical) and the edges are square (and parallel). In the real world, the faces are not parallel, and the cylinder is neither straight nor round. So which height do we report? The height from the low point of face A to the high point of face B; the height from the high point of face A to the low point of face B; or the average? And what do we do if the cylinder is out of round (and trust me, it will be)? Do we report the maximum diameter, the minimum diameter, both, or an average? Part variation is real and cannot be ignored.

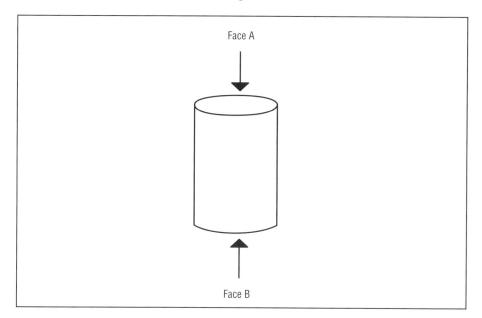

Figure 10-1: Where, exactly, do you measure a part?

Once we begin asking these questions, cages start to rattle. Upper management, and maybe even our own boss, will be horrified by the thought that we can't *"just measure the part and give them an accurate reading."* And the production people will be even more frustrated that we can't *"just tell them if the part is good or bad."*

Most of the time, non-conforming (out of tolerance) parts are produced because some form of variation was misunderstood or even ignored. Taking measurements always involves a degree of classification. If a part is out of round or not parallel, a good measurement system will find a way to incorporate these facts into the inspection report.

Environment – Different shop floor environments can affect a measurement result. Vernier instruments often require the operator to judge which graduation on a scale lines up the best with another line from another scale. If we moved around in our shop to areas of different lighting levels, the lines of the Vernier scales and how they line up with another line may look different. Thermal effects (discussed in Chapter 7) must always be accounted for in measurement planning and designing. If your shop has a stamping department, vibration can have an effect on some if not most of your measurement results. All of these possible and probable environmental effects must be evaluated and quantified in order to control and account for them.

Size of tolerances – Different characteristics will have different tolerances. The smaller the tolerance becomes, the more variation will effect the measurement. When small tolerances are called out and the 10:1 rule (or 5:1 rule) is required, instruments of high resolution are often needed. Instruments are magnification devices. High magnification instruments seem to vary more because of their finer resolution. There's an old saying… "If we could measure all of our parts with a ruler, we'd get great repeatability." If we think of the tolerance size as a target, this becomes crystal clear. If my target is the size of a barn and located 100 feet away, I could hit it in just about any condition, using any firearm. If my target were the size of a CD and located 100 feet away, everything would be a factor: How good is the firearm, what are the winds, etc. Trying to "hit" a small tolerance can require much planning and preparation to hold each factor in check.

When an MSA program is first discussed, it's common for people to make blanket assumptions about measurements. Many times the Gage R&R term will rear its ugly head and you'll hear statements that end with question marks. Such as… well, all we're trying to do is find out how much the gage repeats, right? As if the other five elements (operator, method, part geometry, environment, and size of tolerance) are magically not affecting the measurement.

You should classify measurement systems based on the six elements. When one of the six changes, you have created another unique measurement system. As an example, we'll look at a bore gage measuring inside diameters that are approximately 1" (or 25mm) deep. The walls of the diameter allow the operator to use a good sweeping motion (forward or backward) to measure the part. When the contacts of the bore gage are perpendicular to the walls of the diameter, the indicator reverses direction and the operator can take the reading. Let's assume we have several different part types we are measuring and all diameters are relatively the same nominal size and have the same size of tolerance: we'll use ±.001" (or ±0.025mm). If we look at the six elements of the measurement for each part type, they do not change.

- Instrument: Bore gage with indicator (.0001" or 0.001mm)
- Operator: Standard at the process
- Method: Sweeping method recommended by mfg. instructions
- Part: Nominal sizes are very similar, geometry of the part types are the same
- Environment: All part types are measured at the same shop floor location
- Size of tolerance: ±.001" (or ±0.025mm) on all part types.

We can, with a clear conscience, determine that all of these part types use the same measurement system. But if a new model part has an inside diameter that is only .250" (or 5mm) deep, we have changed the part geometry enough that it will effect the measuring method. Now, when the operator uses the sweeping method recommended by the manufacturer, they don't have near as much surface as the other parts, and the method becomes much more difficult and requires a higher level of finesse. We have changed one of the six elements enough to realistically (with a clear conscience) determine this to be a separate measurement system:

- Instrument: Bore gage with indicator (.0001" or 0.001mm)
- Operator: Standard at the process
- Method: <u>Sweeping method more difficult to perform</u>
- Part: Nominal sizes are very similar, <u>bore is very shallow</u>
- Environment: All part types are measured at the same shop floor location

- Size of tolerance: ±.001" (or ±0.025mm) on all part types.

Another example would be if we had a special part that required a different tolerance. By making a "target" smaller, we have created a unique measurement system:

- Instrument: Bore gage with indicator (.0001" or 0.001mm)
- Operator: Standard at the process
- Method: Sweeping method recommended by mfg. instructions
- Part: Nominal sizes are very similar, geometry of the part types are the same
- Environment: All part types are measured at the same shop floor location
- Size of tolerance: <u>+.0015" (or +0.040mm) on these part types.</u>

When part variation is apparent within a measurement, you'll probably hear the suggestion… "We can just mark the part with a line where we want each operator to measure, right?" Last time I checked, most parts don't come out of a lathe with "measure here" stamped on them. If part variation is present within a measurement, we cannot just mark a fixed point to obtain a better MSA study result. Our goal is to improve the actual shop floor measurement, not "fix" the test.

Another myth I've heard over the years is that you should use secretaries from the office to measure parts during a study (and other absurd wives' tales) to document the worst case operator variation. This is not the case, unless the secretary is sometimes asked to measure actual customer parts. MSAs should be performed using the most realistic recreation of the actual shop floor situation. This will ensure that your studies identify what you do or do not have in a given measurement.

After you determine what you've got within a measurement, it is recommended that you strategically define and refine the specifics of a measurement to control variation. A good example of this is when part variation is inherent to a process. In the measurement method it can be specified that several measurements be taken, and the minimum, maximum, *or* the average reading (whichever applies best within the function of the part) becomes the documented measurement result. If this is specified, it must be adhered to. If the minimum value is specified, judgments (in tolerance or non-conforming) must be made using this value. Often times, people want to have their cake and eat it too by documenting the minimum until a part measures too small (non-conforming) at the minimum, and then they want to pass it because "it's good in most of the other spots I measured." Whatever method is spelled out in the MSA study must be performed in the real-world situation.

Another useful tool involves understanding how each of the elements _should_ react within a measurement. You'll learn which instrument types, because of their design, are more repeatable than others. Micrometers are among the most stable of all precision measuring instruments. If an MSA study is performed using a micrometer, and high levels of repeatability variation show up in the results, it should be viewed as a red flag. I was involved in an MSA project once where a plate thickness was being measured using a 1" micrometer with a resolution of .0001". We noticed unacceptable amounts of repeatability variation, which usually means the parts have variation or the instrument has variation—many times they masquerade as each other. We retested, using a fixed point on the parts, only to determine if the micrometer readings still contained variation. If the repeatability improved, we could determine the variation was from the part variation. If the variation was still present, we could rule out the part and focus on the instrument. The retest yielded very similar results.

The parts themselves had a paper applied to them from a bonding application, and we determined that the contacts from the micrometer were compressing the paper when the measurements were taken. To improve the repeatability to the level expected from a micrometer, we changed from standard contacts to disc type contacts, which disperse the energy of the contacts over a wider area and therefore compressed the paper much less. The retest results yielded better results. It wasn't rocket science; it was just having a fundamental understanding of what the chess moves made with our elements should accomplish (within our measurement results).

Just remember, no matter how much something is universally ignored, it will never disappear. A wise man learns from other's mistakes. We all know the story of how NASA put the Hubble space telescope into orbit and it sent back blurry pictures and two years later astronauts repaired the 60-plus-inch diameter telescope lens with what amounted to a prescription contact lens. The part of the story that is rarely told is that NASA knew the gravity of the Earth caused a "bow" of a few thousandths of an inch in the center of the lens. The engineers also knew the bow would not be present in the "zero" gravity of orbit. NASA engineers could not think of an effective method to measure and/or test the lens in a simulated zero gravity, so you know how they tested it? They didn't!

The elements within your measurement that cause variation will cause variation whether you admit it or not.

Chapter 11
"Know the Uncertainty" is Not an Oxymoron (an Introduction to Uncertainty Concepts and Contributors)

"I always tell the truth… even when I lie…" —Al Pacino's character (Tony Montana) in the 1984 Blockbuster "Scarface"

A phrase that sounds like an oxymoron but actually is not proves to be an extremely powerful metaphor. In Stock Car racing at Daytona and Talladega, two or more cars team up in a draft to get "back to the front." If you've ever been in a casino on the Las Vegas strip and thought you had "even-odds," you really don't know where "the truth lies." **Know the uncertainty** is not an oxymoron.

The purpose of this chapter is to introduce the subject of measurement uncertainty. Every measurement, whether taken in the controlled environment of the inspection laboratory, or in the real-world shop floor environment, is subject to some level of uncertainty. In order to comply with QS-9000 requirements, a facility must understand the concept of uncertainty in measurement and the process of uncertainty estimation.

In most industrial environments, the concept of measurement uncertainty is misunderstood. More often than not, facilities decide what they think measurement uncertainty can be, and blanket that belief over their entire system in hopes they've met the letter and the intent of the QS-9000: 1998, which states:

"Inspection measuring and test equipment shall be used in a manner that ensures the **measurement uncertainty is known** *and consistent with the required measurement capability."*

It comes down to a simple concept of being 1000% certain. If you are 1000% certain, you know what the above requirement means and have tests and documentation in place to offer objective evidence that you are in compliance with this requirement. Very few individuals in industry really understand this requirement. You can often tell how well someone does or does not know a requirement from what their statements end with. "This means uncertainty as in the error documented at calibration… right?" Once again, beware of statements that end with a question mark. If an individual really knows what they're talking about, their statements will end in exclamation points!

First of all, the word calibration is not mentioned in the requirement so this extends far beyond the 15-minute test performed on an instrument every six-months or so. It states: *Inspection measuring and test equipment shall be used in a manner…* This reference is referring to product confirmation, meaning measurements on your shop floor. Secondly, The International Vocabulary of Basic and General Terms in Metrology (VIM) defines *measurement uncertainty* as a "parameter, associated with the result of a measurement, that characterizes the dispersion of the values that could reasonably be attributed to the measurand." In simpler terms, measurement uncertainty is defined as: **The upper limit of how wrong a measurement result can be**. *So let me translate this requirement into plain English:*

> "When inspection, measuring, and test equipment is used on **your** shop floor to confirm the customer characteristics of the parts **you** make for them, the equipment shall be used in a manner in which the upper limit of how wrong each and **every measurement** result can be is known!"

ISO/TS-16949 (the eventual replacement for QS-9000) takes this in an additional direction, when it recommends that internal inspection labs should comply with the requirements of ISO/IEC-17025 (general requirements for the competence of testing and calibration facilities). Before going any further, we should clarify the TS-16949 standard is currently one of the only industrial standards to reference ISO-17025, and does so in the context of lab requirements. This specific reference would not apply to measurements taken on the shop floor. However, QS-9000 states that *"Inspection measuring and test equipment shall be used in a manner that ensures the* **measurement uncertainty is known** *and consistent with the required measurement capability."* We refer to ISO-17025 in this chapter (if for no other reason) because it is an excellent standard to use as a guide when designing and developing an industrial measurement control system.

Conducting uncertainty estimation is quite different from conducting an MSA (Gage R&R) study, which is clearly defined by the MSA reference manual. Uncertainty analysis has a different concept. There is no widely understood manual for referencing the estimating of uncertainty in measurement. **We will never attempt to estimate the exact uncertainty, this would be impossible. Instead, we will establish**

a safe margin of error within our results. This margin of error (k=2 uncertainty) is the range of value(s) that we believe contains the true value (average value of an infinite number of measurements) of the part (see *Figure 11-1*). We should explain this with an example.

Let's assume we measure a plug gage and determine the diameter to be 25.401mm with an uncertainty (k=2) of 0.0025mm. What we are saying is that we're 95% confident that an infinite number of measurements taken on the plug gage diameter would obtain an average value somewhere within ± 0.0025mm of 25.401mm. Let's restate that before proceeding.

- We measure and determine the diameter as 25.401mm with a k=2 uncertainty (95% confidence) of 0.0025mm.

This means that if an infinite numbers of readings are taken of this plug gage, we are 95% certain the infinite average would be within ±0.0025mm of our reading of 25.401mm (the infinite average would be between 25.3985mm and 25.4035mm).

Whenever a measurement is taken, three variables are present (to some degree) in the result: Error, Variation, and Potential Error. To clarify and prove this, we can simply list examples that are present in <u>every</u> dimensional measurement.

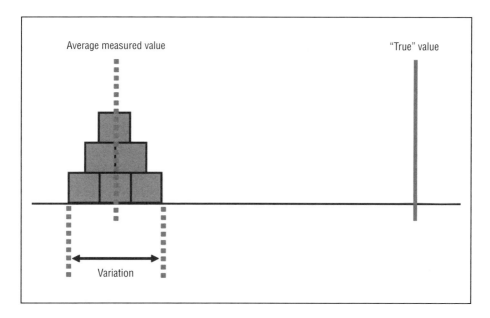

Figure 11-1: Uncertainty: Difference between measured value and "true value".

- **Error:** No instrument is perfect. Every calibration lists an acceptable amount the instrument can deviate from the standard being measured.

- **Variation**: No two objects are identical. Even if they were, we would still get different values when we measured them. In today's real-world resolution of microns and millionths, nothing is truly flat, round, or parallel, and no two operators measure exactly alike. (Note: If variation is not understood to be variation, it can be mistaken for error.)

- **Potential Error**: Unless every measuring lab in the world somehow maintains a temperature of 68.00000°F (20.00000°C) and controls their temperature with the same thermometer, we can always have possible errors when measuring (see Chapter 7).

Now that we understand the examples we can determine concise definitions.

- **Error**: The difference between the measured value and the accepted value of an artifact.

- **Variation**: Difference between two or more elements within a measurement result or results, or the same element during different times or conditions.

- **Error Variation**: The variation of error levels over time or during different conditions.

- **Potential Errors**: Sources of error that may be present within a measurement during any given time.

There are several factors that, when combined, define the specific, complete measurement, as shown in *Figure 11-2*. Each factor can influence the measurement in the form of visible or invisible imperfections. These influences contribute to uncertainty of the result.

- **Measurement setup (reference element)** – In shops where similar parts are produced at different processes, the reference standards (setting masters) used are never <u>exactly</u> alike. This will cause slightly different results when measuring at the different processes. We cannot be certain which result more closely resembles the "true" value.

- **Environment and/or physical constants** – "Unless otherwise specified on an engineering specification, all measurements are based at 68°F / 20°C" (ASME Y14.5-1994 is the standard that governs the Geometric Dimensioning and Tolerancing of your customer's drawings). If environmental effects such as thermal expansion are not evaluated properly, the uncertainty within the measurement can cause incorrect measurement results.

- **Object being measured** – A cylinder shaped object appears round to the naked eye. Several measurements taken using a .00005" (or 0.001mm) instrument will probably obtain slightly different results, making us uncertain of the "true" diameter.

- **Definition of the measured object** – Referring to the same cylinder shaped object in the last example, unless specified, we are not certain if we should record the minimum reading, the maximum reading, or the average? If this is not clearly defined, our result will not be certain.

- **Measurement method** – Quite often, work instructions only specify part characteristic and instrument, but not method. This will enable operators to develop and use different methods to measure the same part. The geometric effects of precision instruments and the difference of each operator's "feel" for the instrument will produce amounts of variation in the measurement results. This lowers the certainty of the results.

- **Measurer(s)** – As you become more involved in conducting measurement training or measurement systems analysis, you will experience how difficult it can be to develop a method where separate operators obtain similar results when measuring in some finer levels of resolution. Measurement values obtained near a tolerance

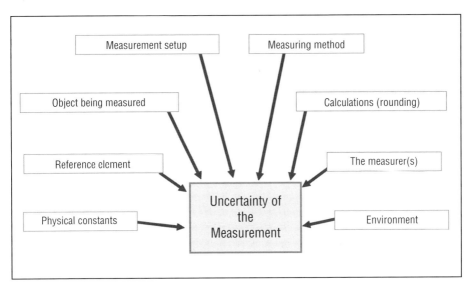

Figure 11-2: Variables within a measurement which can cause uncertainty to occur.

limit are sometimes more skeptical because even the most honest operator in the world will have a tendency to skew to the beneficial side.

The foundation for evaluating uncertainty estimation is the ability to determine the one standard deviation (or one-sigma) of a contributing factor. This concept was explained briefly in Chapter 8. Standard deviation, or sigma, is the measure of range within the readings of a sample (Figure 8-1). In normal distribution, 68% of the readings are within ± one standard deviation from the mean or average value, 95% of the readings are within ± two standard deviations from the mean, and 99.97% of the readings are within ± three standard deviations from the mean.

Among the first things we hear about when learning uncertainty estimation are the "Types" (A and B) of uncertainty contributors. **Type A uncertainty contributors** are statistical tools used to find the standard deviation of a sample group of measurements. Type A evaluations are most helpful because they monitor variation obtained in an actual experiment. A good example of a Type A uncertainty is the result obtained from a robust Average and Range Measurement Systems Analysis study. Type A contributors are often called experimental contributors because they are obtained using a test or experiment. We can justify the estimate of measurement variation using the statistical data from our study. It is always important to remember that a **sample** is just that, and very seldom is a sample detailed enough to display the long-term changes that may be present in a **population**. There are methods of using statistical safety factors (degrees of freedom) to adjust for and safely estimate the difference between a sample and the population, and these will be detailed later.

Type B uncertainty contributors are defined as estimates that are not based on statistical tests. An example of this would be the calibration accuracy specifications of any instrument used. If the accuracy of an indicator is ± .0002", we must be willing to account for this amount of **potential error**, because we cannot eliminate it by any formula or equation. Furthermore, even if we confirm the indicator to routinely possess less error (or higher accuracy), we cannot assume this higher level of accuracy will be constant or present at all times. Type B uncertainty contributors are often called "estimated contributors" because they are not derived from a test or experiment. They provide the latitude to use information such as manufacturer specifications, as-left error documents on calibration certificates, or knowledge of inherent error(s) from the mechanical workings of the instrument, such as hysteresis within an indicator measurement, to estimate uncertainty effectively. Since Type B contributors are not derived from statistical analysis, normal distribution cannot be assumed. We will apply other distribution factors. As each uncertainty contributor is determined it can be classified as Type A (which assumes normal distribution) or Type B.

As uncertainty contributors are identified and evaluated, we will need to convert the effect of each contributor into the same unit of measurement or "currency" in order to combine them. Using different distribution factors helps perform this conversion. Each distribution factor is used to evaluate a different type of phenomena.

- **Normal distribution** is used for Type A uncertainty contributors (*Figure 11-3*).

- **Triangular distribution** is most often used in evaluations of noise and vibration. These "hard" limits are easier to estimate than sigma (*Figure 11-4*).

- **Rectangular distribution** is conservative. This is often used with error from calibration certificates. We know the limits of variation, but have little idea about the distribution of uncertainty contributors between the limits (*Figure 11-5*).

- **U-shaped distribution** is not as rare as it seems. Cyclic events, such as temperature, often yield uncertainty contributors that fall into this sine wave pattern (*Figure 11-6*).

Figure 11-3: Normal distribution.

Figure 11-4: Triangular distribution.

Figure 11-5: Rectangular distribution.

Figure 11-6: U-shaped distribution.

To comprehend how these distribution factors work with different uncertainty contributors, in the next chapter we will take a brief look at some of the common uncertainty contributors or types of contributors in measuring. The examples used are only samples to give a brief depiction of the population of uncertainty contributors in today's industrial world. The examples are not inclusive; your facility may encounter contributors that differ because of the specific application or environment of the measurement.

There is no overstating how important it is to correctly estimate the uncertainty within the measurement due to thermal effects. These thermal effect scenarios represent a fundamental rule of uncertainty estimation. **Our goal is not to estimate the exact uncertainty—this would be impossible—but to estimate each contributor safely enough so that we can satisfy any argument from a customer, competitor, or auditor if they feel we have underestimated.** It's the same concept used for years when estimating how much time to allow for travel. We don't want to guess exactly how much time it will take; we just want to estimate safely to allow enough time.

Chapter 12
Uncertainty Part Two: Budgets and Estimation Concepts

"Probability is the clock that God gave mankind to tell the time of the World"—Gonzalo García-Pelayo

Both Type A and Type B uncertainty contributors are needed for estimation when applied in the correct situation. Uncertainty is an engineering estimation based on data or a documented phenomenon. These can usually be explained as approaches to how we estimate. If we needed to determine the one standard deviation of the speed of cars traveling past a designated spot on a street in our community which had a posted speed limit of 35 MPH, we could make this estimation using Type A and Type B approaches.

The Type A Approach

1. Buy a radar gun.

2. Spend several days hiding behind a lamppost (at different times of the day and night).

3. Calculate the standard deviation of the data.

4. This standard deviation, or distribution, would be based on data obtained from a sample. We would use it to estimate the standard deviation of the population, or infinite number of cars. If we conducted the study several times, our data would be slightly different each time.

The Type B Approach

1. Drive down the street several times.

2. Compare our speed with the fastest "maniac" and the slowest "sight seeing driver."

3. Assume normal distribution, calculate the standard deviation.

In many situations, your experience with the contributor will allow you to make a safe estimate in a fraction of the time, effort, or cost. You'll learn when and to what level you can trust your experience.

I haven't found any area of uncertainty estimation where this experience comes into play more than when estimating the uncertainty introduced by thermal effects that may be present within a measurement. Correctly estimating thermal effects begins with understanding the formula to determine the thermal expansion of a specific material after a given change in core temperature. Before you begin to estimate the measurement uncertainty from thermal effects, you may want to review Chapter 7.

Unfortunately, it is seldom this simple when we have to estimate the uncertainty within a measurement due to thermal effects. In many measurements, the uncertainty from thermal expansion not only effects the part being measured, but also the instrument used for measuring and the dedicated setting master (if used) to reference the measurement. When these factors exist, we begin dealing with uncertainty correlation issues.

This happens because of the different expansion scenarios created within measurement setups. In some measurements, the instrument and part expand in the same direction. This means the uncertainty caused by the thermal expansion of both is less than with the expansion of the part alone. In other measurements, the part and the instrument expand in different directions. This creates the possibility of uncertainty caused by the thermal expansion of both, which may be greater than the expansion of the part alone.

In the caliper measurement shown in *Figure 12-1*, our steel part measured 49.28mm. If the part and instrument were moved to a shop floor environment of 85°F, we could assume a thermal effect on the steel part and on the jaws of the caliper. The expansion on the jaws would be very small and with the caliper resolution of 0.01mm would not likely be detected. If it were, it could be factored out of the equation when zeroing the LCD display on the caliper before measuring. We could then safely estimate the thermal effect of the part, and the thermal effect on the instrument could be considered insignificant in this application and circumstance.

Another scenario is when a steel part is brought into a lab for measuring. If the part has not been in the lab long enough to stabilize to the lab environment, there will be uncertainty within the measurement due to the **average temperature difference** between the measuring equipment and the part being measured *(Figure 12-2)*. The average temperature difference between the shop environment and the lab environment (where the gage blocks are used as a master) is 11°F (6.11°C). The

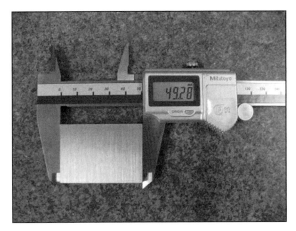

Figure 12-1: Any thermal effect on the instrument contacts would be factored out when the LCD was zero-set. Image courtesy of Mitutoyo America and Qualtech Tool & Engineering.

equation to determine the thermal effect in this measurement would look like this:

Change in Length =

Original length × CTE × Temperature difference (between gage block used as a master, and the actual part)

.87365" (Reference length of part) × 0.0000064 × 11°F

Change in Length = .000062"

Metric equivalent:

22.1907mm × 0.0000115 × 6.11°C

Change in Length = 0.001599mm

Author's note: Because the conversion between °F and °C is a repeating decimal (1°F = 0.55555...5°C), trying to convert metric and inch thermal expansion scenarios will never result in an <u>exact</u> conversion. Always search for an exact conversion if possible. For example, to convert inch to metric, always multiply by 25.4 because this conversion is exact. (1" = 25.4mm). Using the conversion (inch to metric) of dividing inches by 0.03937008 is not exact.

All thermal effect uncertainty contributors follow u-shaped distribution and should be multiplied by a distribution factor of 0.7.

The most common Type A uncertainty will usually be the variation present within the measurement itself. **The best way to determine this variation is by performing an MSA study that conforms to the criteria found in the AIAG MSA reference**

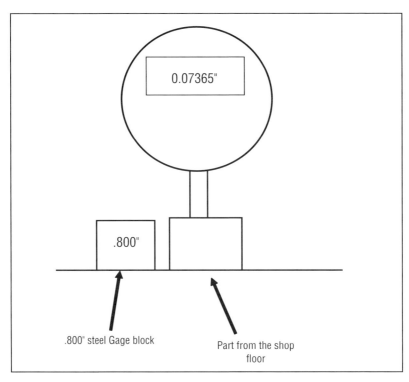

Figure 12-2: There will be a temperature difference between gage blocks in the lab and parts from the shop floor.

manual. It is safe to view the MSA study result as determining 5.15 sigma, or approximately 99%, of the measurement variation. In order to estimate the uncertainty from the MSA variation, we will need to calculate the one-sigma process variation and then determine the effective degrees of freedom. A conversion table can be used to estimate the degrees of freedom. The equation **n – 1** is used where **n** is equal to the number of samples in the population. This is compared to the table to determine the statistical safety factor. This is sometimes called a "standard deviation of the standard deviation."

Sample size:	1	5	10	20	50	∞
Safety factor:	6.99	1.33	1.14	1.07	1.03	1.00

Example

We conduct a Gage R&R study on the measurement of a part that has a print tolerance of ± 0.1mm, using 10 sample parts, three appraisers, and three trials (90 total readings). The variation consumes 17% of the 0.2mm print tolerance, or 0.034mm.

Uncertainty = one sigma

= (0.034 / 5.15 sigma)

= 0.0066 × **1.03** (Statistical safety factor for our sample size)

= 0.0068mm

Since variation is a Type A contributor, normal distribution is assumed.

Another uncertainty contributor occurs when using digital instruments. These instruments use magnetic strips or glass scales that a display "reads" to determine the amount of movement the instrument has made. As the gage moves, the display counts the increments or units of resolution. The uncertainty occurs because when the gage moves half-way between steps, the display adds another increment.

Example

An indicator (which was zero-set using a .700" ceramic gage block) has a LCD resolution of 0.001mm or .00005", and as the plunger is moved upward the display counts the digital steps. Each time the plunger raises past half-way of the digital step, the LCD display adds an increment of 0.001mm (or .00005"). The uncertainty occurs because, if the true value of the part is any measured value between .773626" and .773674", the display would read .77365". Whichever reading we are using for our measurement, we can safely estimate a digital step uncertainty contributor of one-resolution of the digital instrument—in this case, 0.001mm or .00005". For this uncertainty contributor, a digital step distribution factor of 0.3 is assumed *(Figure 12-3)*.

To help understand this concept of uncertainty, keep in mind that it is similar to, yet also different from, the digital step concept. Let me explain. If I handed you a cell phone and gave you one-second to look at the time it displayed (and we'll say it displayed 1:32PM), how close could you estimate the <u>exact</u> time down to the second? Think about it, the actual time could be anywhere between 1:32:00PM and 1:32:59PM, yet the cell phone would display 1:32. Your uncertainty would be 59 seconds. If you used the digital step concept and estimated the exact time to be 1:32:30 you would cut your uncertainty in half. Even though the cell phone did not display the time in seconds, you could guarantee a safer estimate by taking the middle. If the exact time was 1:32:00, you'd have estimated within 30 seconds, and if the exact time was 1:32:59 you would have estimated within 29 seconds. Digital step is similar (in concept) to the time uncertainty example. It all comes down to knowing where the

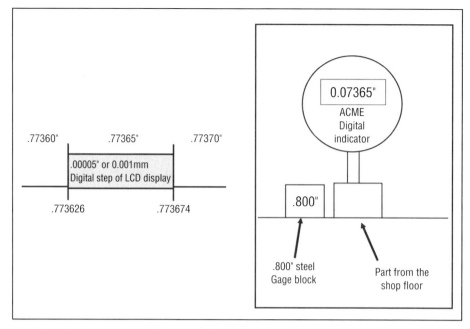

Figure 12-3: Uncertainty from the digital step of the instrument display.

display change point is. The cell phone changes after a complete minute and most digital instruments change halfway between two-increments.

Information concerning the error(s) inherent to the measurement instrument itself can be found in the manufacturer's literature received with the instrument. Usually, this information can also be obtained by contacting the instrument manufacturers. Some will provide this material free of charge, and since most of the instruments are made using the same mechanical principles, it is easy to develop a matrix of possible errors for each instrument type. An example could be found in the owner's pamphlet from a maker of precision measuring tools, which lists the following errors as being inherent to the construction of a Vernier caliper.

- Parallax error – One-graduation.

- Abbe's error – We could calculate the formula listed if we had a Master's Degree in Calculus. So, we'll just use one-graduation.

- Resolution error – Two people may always view a Vernier scale differently, so that's one-graduation.

These error factors should be estimated separately. If and when they apply to the measurement, digital step or rectangular distribution is assumed, depending on the contributor.

Another uncertainty contributor is the calibrated accuracy (or allowable inaccuracy) of an instrument. This information can be found on the certificate of calibration, and it should be assumed for every instrument or artifact used in the measurement. For example, if we were using a micrometer to take a direct measurement of a part, we would use the calibration accuracy of the micrometer as an uncertainty contributor. If we were using a gage block to zero-set a micrometer and then measuring a part to obtain a value referenced to the gage block, we would need to use the accuracy of the gage block and the micrometer as separate uncertainty contributors. Rectangular distribution is assumed for these uncertainty contributors because only the limits are known.

Root Sums Squared

The first step in evaluating the actual measurement is to identify all of the uncertainty contributors and separate them into Type A and Type B contributors. The effect of each contributor is multiplied by the appropriate distribution factor. This is now termed the "**sum**" of each contributor. The sums are squared and added together. The square root of the sums squared becomes the "**combined uncertainty**," which is multiplied by a k factor—we will be using k=2—to determine the **expanded uncertainty**.

Now we can begin to examine our measurements to estimate the uncertainty of each contributor. We will identify and quantify each contributor, and then multiply the result by the appropriate distribution factor. The following example is used by courtesy of HN Metrology Consulting in Indianapolis, IN.

Example: ID/OD Comparator Measurement

An 85mm ring gage is calibrated on an ID/OD comparator. A stack of gage blocks of the same nominal size as the ring is used to "master" the comparator.

- Measurement Conditions

1. Temperature in the laboratory is 19.5° to 20.5°C.

2. The temperature difference between the master gage block stack and the measuring object is estimated to be less than 0.1°C.

3. The part is made of steel.

4. The master gage blocks are made of steel.

5. The operator is trained and familiar with the wringing of gage blocks and the use of the comparator.

- Uncertainty Contributors

1. Difference in expansion due to difference in thermal expansion coefficient (CTE).

2. Difference in expansion due to difference in temperature.

3. Comparator resolution.

4. Uncertainty of length of master gage block stack.

5. Repeatability including the effect of wringing.

- Quantifying Uncertainty Contributors

Table 12-1 summarizes the uncertainty contributors.

Now we can use the square root of sums squared formula to combine the uncertainty.

Combined standard uncertainty = Square root of ($S1^2 + S2^2 + S3^2 + S4^2 + S5^2$)

Combined standard uncertainty =

Square root of (0.035μm + 0.065μm + 0.003μm + 0.06μm + 0.04μm)

Combined standard uncertainty = 0.10μm.

Contributor	A/B	Distribution	Number of Measurements	Variation Limit (+/-)	Standard Uncertainty
Difference in CTE	B	U-shaped	-	0.05μm	0.035μm
Difference in temperature	B	U-shaped	-	0.09μm	0.065μm
Resolution	B	Resolution	-	0.01μm	0.003μm
Master Uncertainty	B	Normal	-	0.12μm	0.06μm
Repeatability	A	-	30	-	0.04μm

Table 12-1: Standard uncertainty. Image courtesy of HN Metrology Consulting.

To expand the uncertainty we multiply the combined uncertainty by a factor of k=2

Uncertainty = 0.10µm × 2 = 0.20µm

In the words of Dr. Henrick Nielsen "and that's all there is to it." You can figure the Root Sums Squared method of uncertainty calculation on a $5 calculator.

Something to keep in mind as you plan your uncertainty studies and develop your uncertainty budgets is how different approaches can be used to account for uncertainty contributors. Some organizations are very comfortable with using a basic MSA study to determine the Type A contributors, and then they do very detailed estimates of the Type B contributors individually. Some facilities develop very robust MSA studies to account for many standard Type B contributors in their MSA (Type A) data. For example, in our uncertainty budget we could have collected MSA data from measurements taken on different shifts during a twelve-consecutive-month time frame. If this type of "robust" MSA were performed, we could safely estimate the uncertainty from the thermal effects (within the measurement) to be accounted for in the variation of the measurement data from our MSA results. This concept includes traditional Type B contributors in the Type A estimate. Either approach is acceptable as long as the contributors are safely estimated. As your facility becomes more comfortable with developing uncertainty budgets, you'll determine which type of study is best for different measurement applications.

Always remember: Your goal is to safely estimate all uncertainty contributors. The rule of thumb is that if you cannot prove a given phenomenon is <u>not</u> a contributor, you've probably proven that it might be one. And if it might be… include it, just to be safe.

Chapter 13
Selecting and Implementing Software

*"You can plan a pretty picnic,
but you can't predict the weather..."*—Outkast

As your facility prepares for ISO-9000, QS-9000, or TS-16949 registration, or strives to meet the increasing demands of your quality system's processes, you must demonstrate increasingly better control of inspection, measuring, and test equipment. Gage management software is considered by many to be the best tool for ensuring this control. Deciding which version of software will meet a facility's needs and requirements can seem like a shot in the dark, as it can be difficult to envision what the software can do for your system.

Imagine trying to decide on a new car purchase, having never driven a car. Add to this the reality that once the bank (management) cuts the check for the car (the software), it better get you where you need to go (in search of the ultimate gage management system), because it may be five- to ten-years before you're considered for another bank loan (software upgrade). First of all, you must understand and accept that software itself will not make or break a system. The two important questions to consider are:

1. What does the software allow the user to do?

2. What can the user's creativity and flexibility do with the software?

Each question represents separate *perspectives* of a philosophy. What you get out of the software is a shuffled version of what you put into the software. The best course to take when you "test drive" a demo version of gage management software is to follow a sample instrument's life cycle. This cycle is a loop of gage data, calibration data, and issue records that will be discussed in more detail in the next chapter. The software must document this loop and be user-friendly for the purposes of proving its veracity. The first format you must consider is the initial gage record itself. Think of this as the "home page" for each individual device of inspection, measurement, and test equipment. Different software packages have different terms for this section. We'll call this the DEVICE INVENTORY section. In this section we'll need to know some basic information about the type of instrument, basic history, and some information about its current status within the system. You can see this in the example instrument detailed in *Figure 13-1*. When we pull up this device inventory page within our software database, we find that this instrument is an Ametek Pressure Gage that is located in the lab and assigned to a Supervisor. The gage is also IN (within) established accuracy specifications, can be used as a master, and is due for calibration on the listed date. This software window also allows us the latitude to record some specific information about the gage and/or its history. This is a very beneficial balance within gage software: having many preloaded drop-downs and other features to allow fast data entry, and it also provides the opportunity to record detailed comments that will serve as very useful documentation later. If this seems like a lot of information that has nothing to do with calibration, always remember we are looking for much more than "calibration" software. <u>We want an inspection, measurement, and test equipment management system</u>.

After our initial information is entered into the database, we can move on to the calibration itself.

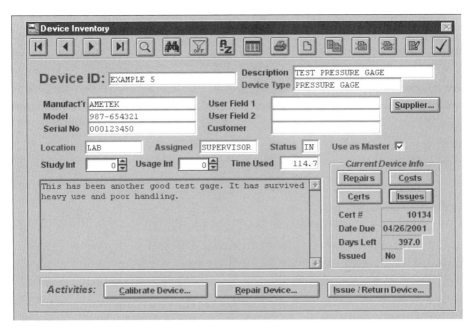

Figure 13-1: The general information about each instrument or device. Image courtesy of GMS for Windows by Prompt Consultants, Inc.

CALIBRATION INFORMATION: This section should provide more flexibility. The interface will probably look much like a calibration certificate from an accredited lab and contain much of the same information. A very useful feature within this interface is the ability to assign specific test characteristics to individual instruments. As you notice in the example record seen in *Figure 13-2*, when the record is called up we see everything we need to see without scrolling. This is a great time saving feature. Not only do you not have to scroll to find what you need, but you will not need to scroll to find what your auditor or your customer may need to see. We can tell right away what the status of this calibration is, what characteristics were measured, what the measured results were, and what the environmental conditions were. We also have a shortcut to the maintenance records and the calibration procedure for this instrument. And, again, we have the latitude to enter very specific information about this specific calibration. In this case, we used this option to list the standards used for calibration.

Another useful feature that I look for in gage software is compatibility between its display and interfaces with current accuracy standards. For example, in Chapter

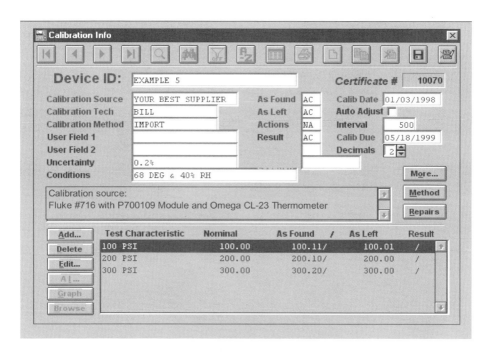

Figure 13-2: The specific information about each calibration or test. Image courtesy of GMS for Windows by Prompt Consultants, Inc.

Chapter 13: Selecting and Implementing Software   115

4 we discussed accuracy classes for cylindrical gages such as plug gages and ring gages. These tolerances can be seen in *Figure 13-3*.

This accuracy standard actually specifies six-measurement-points for a plug or ring gage. The gage must be within the listed accuracy along its X- and Y-axis at the top of the gage, the middle of the gage, and the bottom of the gage. In our example software, we have the preloaded ability to set this up within our calibration information record, as can be seen in *Figure 13-4*.

The life-cycle of an instrument includes purchase, initial calibration, and its being issued to the specific location where it will be used. Your software should have a very

Gagemaker's Tolerance Chart							
Above	To & Included	CL - XXX	CL - XX	CL - X	CL - Y	CL - Z	
0.010" .254mm	0.825" 20.95mm	.000010 0.25μm	.000020 0.5μm	.000040 1.0μm	.000070 1.75μm	.0001 2.5μm	
0.825" 20.95mm	1.510" 38.35mm	.000015 0.38μm	.000030 0.75μm	.000060 1.5μm	.000090 2.25μm	.00012 3.0μm	
1.510" 38.35mm	2.510" 63.75mm	.000020 0.50μm	.000040 1.0μm	.000080 2.0μm	.00012 3.0μm	.00016 4.0μm	
2.510" 63.75mm	4.510" 114.55mm	.000025 0.63μm	.000050 1.25μm	.0001 2.5μm	.00015 3.75μm	.00020 5.0μm	
4.510" 114.5mm	6.510" 165.35mm	.000033 0.83μm	.000065 1.625μm	.00013 3.25μm	.00019 4.75μm	.00025 6.25μm	
6.510" 165.35mm	9.010" 228.85mm	.000040 1.00μm	.000080 2.0μm	.00016 4.0μm	.00024 6.0μm	.00032 8.0μm	
9.010" 228.85mm	12.010" 305.05mm	.000050 1.25μm	.0001 2.5μm	.0002 5.0μm	.0003 7.5μm	.0004 10μm	
12.010" 305.05mm	15.010" 381.25mm	.000075 1.88μm	.00015 3.75μm	.0003 7.5μm	.00045 11.25μm	.0006 15μm	
15.010" 381.25mm	18.010" 457.45mm	.0001 2.50μm	.0002 5μm	.0004 10μm	.0006 15μm	.00080 20μm	
18.010" 457.45mm	21.010" 533.65mm	.000125 3.13μm	.00025 6.25μm	.0005 12.5μm	.00075 18.75μm	.001 25μm	
21.010" 533.65mm	24.010" 609.85mm		.0003 7.5μm	.0006 15μm	.0009 23μm	.0012 30μm	
24.010" 609.85mm	27.010" 686.05mm		.00035 8.9μm	.0007 18μm	.00105 27μm	.0014 36μm	
27.010" 686.05mm	30.010" 762.25mm		.0004 10μm	.0008 20μm	.0012 30μm	.0016 41μm	

Figure 13-3: National Accuracy standards for cylindrical gages. Image courtesy of Glastonbury Southern Gage.

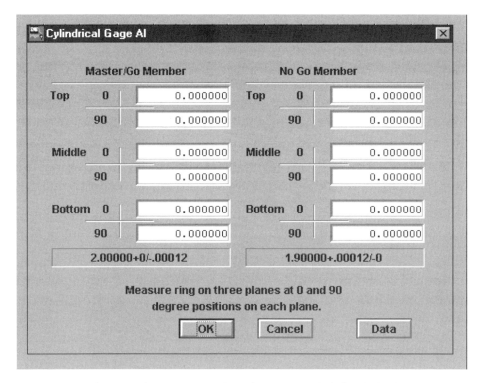

Figure 13-4: The software interface "matches" the National Accuracy standards for cylindrical gages. Image courtesy of GMS for Windows by Prompt Consultants, Inc.

easy issue and return feature to quickly assign or reassign instruments as the needs of the facility change (*Figure 13-5*).

In some software, this feature can also offer a scanning system option in which barcode labels can be attached to instruments so transactions can be made and recorded at the sweep of a wand. Not only is this much faster, it is also virtually mistake-proof, which is very beneficial as (you will no doubt find) typo errors will likely be the most common "virus" within the database of your measurement control system.

After gage management software is installed and implemented, the instrument data within the database will have to be retrieved for all of the reasons we can automatically think of, but, more importantly, for the many others that we cannot think of. Regardless of why we need the data and reports, software is only as good as its ability to offer multiple "filter" options when selecting and reporting. A "report manager" is shown in *Figure 13-6*.

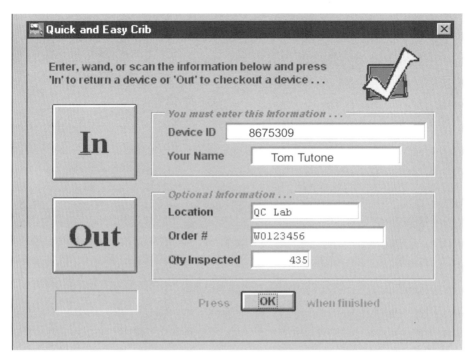

Figure 13-5: An easy issue and return feature is very desirable. Image courtesy of GMS for Windows by Prompt Consultants, Inc.

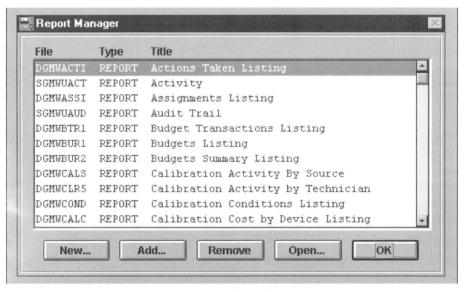

Figure 13-6: Pick a report, any report... Image courtesy of GMS for Windows by Prompt Consultants, Inc.

Planning and organizing determines how well the software manages the system. I suggest developing a matrix to separate the different types of instruments you have in your inventory. Most of the categories will be simple. The not-so-common instruments can be placed in one or more miscellaneous categories. Each category of instrument can be examined for certain generic needs. Calibration interval, Vernier or digital, is it used on the shop floor or the inspection lab? You can now look at **every window** in the software interface and determine how you want each window to read. This is your chance to be creative. Take a look at outside source calibration certificates, and model the software system after the documents and formats that impress you the most. Follow the cycle we discussed earlier and set up each line and window as perfectly as you can. When this is completed for a gage type, you can enter one gage of this type in the system. This gage will now become the "copy" for the gage type. Some software packages allow copying a new gage exactly like the copy gage. For obvious reasons, the calibration data may not be copied because this data is unique to each gage. Still, you can use this concept because you can print and retain the calibration record of the copy gage. All Windows©- based software has a copy/paste feature where you can call up the window of the copy gage and copy everything in the window to the same window of another gage. This capability is priceless.

Here's how it works at real-world speed. All the printed information from the copy gages can be kept in a binder for reference. A list of all of the separate gage types and identified copy gages can be placed somewhere near the PC. (I have my list laminated and taped to the bottom of my monitor, right above my keyboard.) When calibration data are being entered, the user can simply call up the copy gage, copy the information, and paste where needed. <u>If this is done properly, every gage within its category is documented in exactly the same way.</u>

Another useful tool to look for in gage software is some sort of supplier or provider database (*Figure 13-7*). You're only as good as the people and facilities you have on speed-dial. If you're looking up information about an instrument during a critical time, and need the assistance of the facility that you purchased it from, their contact information should only be a click away.

As your system matures with the use of software, you'll find yourself acting like a NASCAR crew chief; you'll always find ways to make things just a little better. As you incorporate improvements and the information (being entered) changes, updates can be printed for your reference binder. The calibration interval for each copy gage will tell you when the improvement will "cycle through" the entire system. There is no bigger thrill during an audit than to walk the floor with an auditor and gather Gage

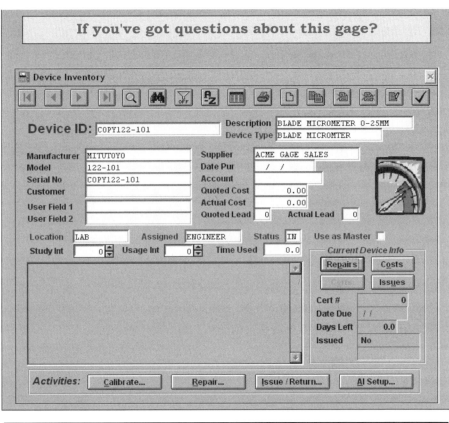

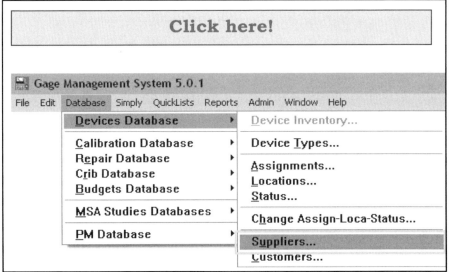

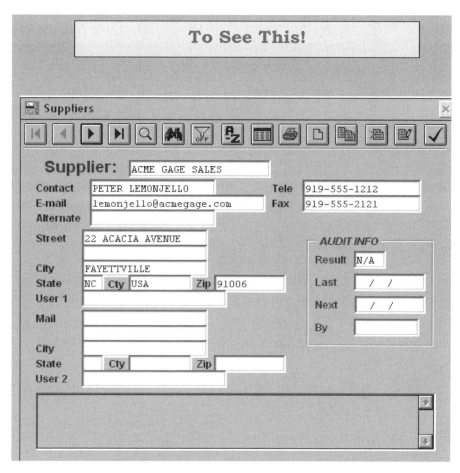

Figure 13-7: Links to OEMs and/or suppliers. Image courtesy of GMS for Windows by Prompt Consultants, Inc.

ID numbers of 75 to 100 gages. Then, in the lab, the auditor gives up after checking only 15 or 20 calibration records. It can be done.

Remember: Audit your system constantly. As mistake-free as the copy/paste system is, you will always have people that think they don't need to use it. I've seen information typed in wrong and then copied over to several more records. When this starts happening, it will spread like a virus.

You'll know you have a great system when somebody calls you from a conference room asking for some information during a major scrap or rework dilemma at your facility (the gage is always the first suspect in a measurement problem). Forty-five seconds later, when you print it out and place it in the person's hand, the emergency just drifts away toward someone who can't say what they did and can't back up their actions with documentation. Later, somebody will ask you what the crisis was all about and you'll smile and then try to look serious as you reply: "I don't know. After I gave them my information, I never heard from them again."

This will eventually become your rule to which everyone else will take exception.

Chapter 14
Developing a Smooth Flow

Knowing is not enough, we must apply…
Willing is not enough, we must do… —Bruce Lee

A world-class industrial precision measurement control system is a complex machine that, when properly applied, represents a model of mathematical precision. Like any complex machine, it uses hundreds and possibly thousands of moving parts that all work in unison to, at times, perform as eloquently as a symphony composed by Mozart and, at other times, perform as emotionally raw as Jimi Hendrix playing the Star-Spangled-Banner.

Measurement instruments, and all activities they are used for, must be documented from the cradle to the grave. The most effective way to create this is to establish "clone" or "copy" gages within your database. These copy gages can be referenced quickly to serve as a model within the model that all instruments must emulate. For example, let's assume on our shop floor we have 58 micrometer instruments and, of those 58, we can break them down to fit nine different or separate applications:

Blade micrometer	Disc micrometer
Inside micrometer	Depth micrometer
Outside micrometer	Spline micrometer
Point micrometer	Internal bore micrometer.
Ball micrometer	

Following this matrix, we can break down each application category into subgroups. For example, of our 15 blade micrometers, we have:

- Three Mitutoyo model 122-101

    0-25mm/0.01mm

    ±0.003mm accuracy

- Two Mitutoyo model 122-102

    25-50mm/0.01mm

    ±0.003mm accuracy

- Two Mitutoyo model 122-103

    50-75mm/0.01mm

    ±0.003mm accuracy

- Two Mitutoyo model 122-125

    0-1"/0.001"

    ±.00015" accuracy

- Two Mitutoyo model 122-126

    1-2"/0.001"

    ±.00015" accuracy

- Two Mitutoyo model 122-127

    2-3"/0.001"

    ±.00015" accuracy

- One Mitutoyo model 422-311-10

    0-1", 0-25.4mm/0.00005", 0.001mm

    0.001mm/0.00015" accuracy

- One Mitutoyo model 422-312-10

    1-2", 25.4-50.8mm /0.00005", 0.001mm

    0.001mm/0.00015" accuracy.

Now we can document one instrument from each category as our clone or copy gage.

- Mitutoyo model 122-101: Unique gage ID #8675309
- Mitutoyo model 122-102: Unique gage ID #2245386
- Mitutoyo model 122-103: Unique gage ID #529761

- Mitutoyo model 122-125: Unique gage ID #4466789
- Mitutoyo model 122-126: Unique gage ID #567284
- Mitutoyo model 122-127: Unique gage ID #224689
- Mitutoyo model 422-311-10: Unique gage ID #5542189
- Mitutoyo model 422-312-10: Unique gage ID #567821.

Another concept is to create "phantom" instruments within your database. Phantom instruments are "dummy" gages within your database that don't really exist (make sure your software will allow you to assign them this way), but are added to your database to copy data from. There are two advantages to doing this. First, your copy gage is a phantom, so it can never be lost or unaccounted for, and secondly, you can assign unique gage ID numbers that are very easy to recall, such as:

- Mitutoyo model 122-101: Unique gage ID #Copy122-101
- Mitutoyo model 122-102: Unique gage ID #Copy122-102
- Mitutoyo model 122-103: Unique gage ID #Copy122-103.

By doing this, we have created phantom gages to use as a copy gage, and their unique gage ID numbers are #Copy + the model number (*Figure 14-1* and *Figure 14-2*). This system is very easy to understand and explain. I'm sure I didn't invent it.

We can enter the copy gages into our database to include all common characteristics. By common, we mean common between more than one instrument. A specific characteristic, such as a calibration record, will be specific to a unique instrument, and we would not want to be copying this information into another instrument's record within our database.

After an instrument is copied into the database it will need to be calibrated. Your facility will have controlled instructions for calibration of each family of instruments as suggested in Chapter 3. If you find the right software, these procedures will be loaded into the software for your reference any time you need to review them. These electronic documents can become controlled at your facility, because the instructions cannot be modified by the software user. I would suggest printing a copy of each instruction for your facility's document control system records. The controlled document could reference the version of the software as the source of the document. As your software is upgraded from time-to-time, I would suggest reviewing the calibration instructions loaded into the upgrade and, if there are any changes, the con-

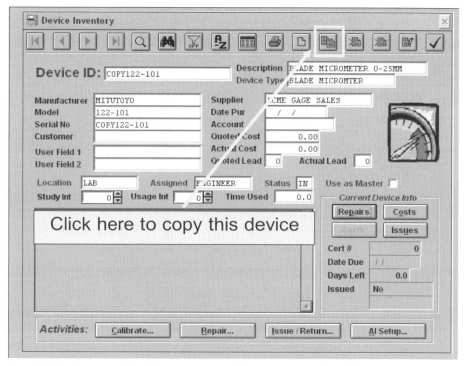

Figure 14-1: Copying a gage should be as easy as one click. Image courtesy of GMS for Windows by Prompt Consultants, Inc.

trolled copies of the instructions can be revised to reflect the changes in the upgrade. In the example in *Figure 14-3*, we would reference our calibration instructions to Gage Tolerance Micropedia Pro 4.0.

Now we can enter a calibration record, which will become the backbone of your instrument control system. I would highly recommend that calibration data be entered, checked, and confirmed to make sure it documents exactly what you expect it to. When choosing software, you'll usually find two options for calibration record interfaces. One option has separate tabs for the situation, environment, data, etc. This is very comprehensive, which has its advantages. The second option is more basic and easier to manage, confirm, and audit. It condenses the most important documented information about the calibration into one easy to explain and easy to understand interface. During audits of the Element 4.11 system, it can be confusing to an outside auditor if you have to "tab" through several interfaces to explain the documentation of a single calibration. This thought process is based on a simple fact; software does

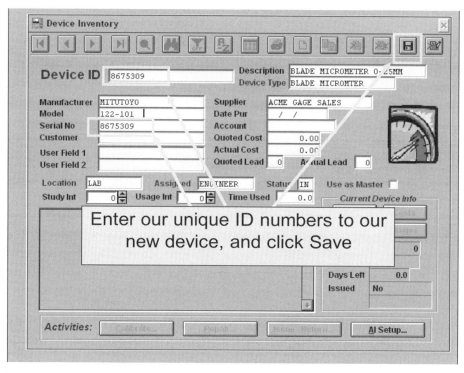

Figure 14-2: After copying, the new gage can be easily saved. Image courtesy of GMS for Windows by Prompt Consultants, Inc.

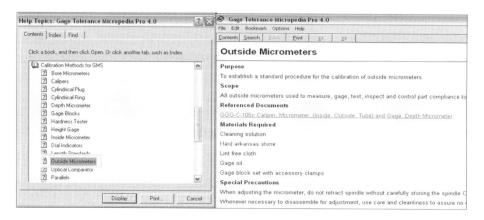

Figure 14-3: Calibration procedures within the software. Image courtesy of GMS for Windows by Prompt Consultants, Inc.

not make or break your system, it magnifies what makes up your system. It can serve as a tool to magnify strengths or expose weaknesses.

If you look at *Figure 14-4* of the calibration record from our example software (GMS for Windows V5), you can easily understand and explain exactly what the record documents. An internal calibration was performed on 5/23/05 on Micrometer #8675309. The instrument was tested by taking length measurements at increments of 5.100mm, 10.200mm, 15.300mm, 20.400mm, and 25.000mm. The temperature during calibration was 68.4°F and the maximum error at any given test point was 0.002mm. The instrument is due for recalibration by 11/22/05. Even though you are using a compact interface, all the information you may be asked about is only a click away.

After a while, you'll learn to anticipate the questions commonly asked by auditors or customers, and good software is built to answer these questions quickly. For example, if an auditor was reviewing the calibration record for Micrometer #8675309, he or she is almost guaranteed to ask about the documented procedure for the internal

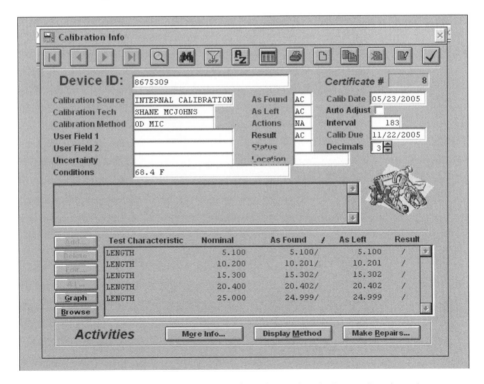

Figure 14-4: The calibration details. Image courtesy of GMS for Windows by Prompt Consultants, Inc.

calibration. For the answer, we can just click "Display Method," and the interface takes us directly to the calibration procedures loaded within the software (*Figure 14-5*).

The next question most likely to be asked is something along the lines of "what calibration standards were used during this calibration." If we click "More Info," the software brings up the standards used that match this specific procedure (*Figure 14-6*). This is what developing a flow within your system is all about: having all the tools in place, and arranging them in a way that allows them to be used in one move.

After a new instrument is copied into our system and calibrated as prescribed by your internal standards, you will need to issue, or as I like to say "deploy," the instrument to the desired location for use on the shop floor. In our sample software, this is as simple as clicking "Issue/Return," and you can deploy the instrument as needed (*Figure 14-7*).

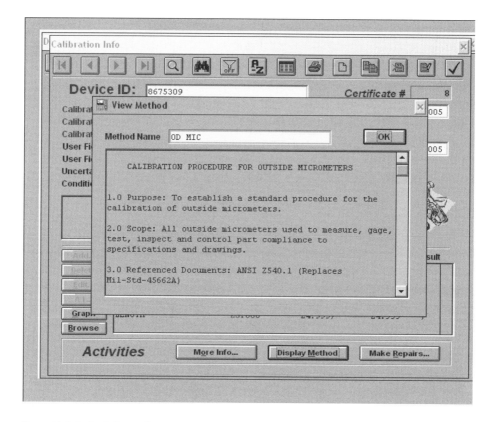

Figure 14-5: Preloaded procedures answer any questions about the calibration process. Image courtesy of GMS for Windows by Prompt Consultants, Inc.

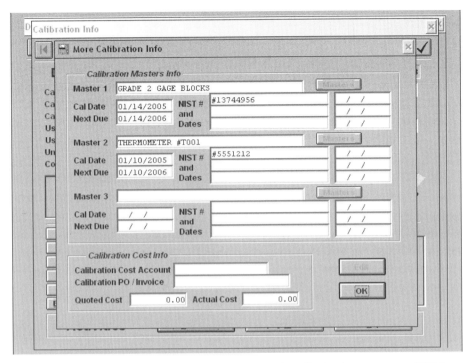

Figure 14-6: Links directly to the artifact standards used during the calibration. Image courtesy of GMS for Windows by Prompt Consultants, Inc.

This clearly demonstrates how, with the correct tools, you can place yourself at "Gage Command Central." As you are issuing the instrument, you receive "real time" data and information as to the issue date and time, as well as how much calendar time remains before the instrument is due for recalibration. After the issue information is entered, you can save this issue record (*Figure 14-8*).

This becomes the bread and butter of a system: The cycle of receiving an instrument, entering the instrument's details into the software, calibrating the instrument, issuing the instrument, and retrieving the instrument for recalibration. Occasionally, the instrument may leave the cycle to be repaired outside your facility, and the software can still keep track of its whereabouts. Software is the most important tool at your disposal to document the flow.

Like anything else in life, you really don't know how well a system is until it collides with some real-world resistance. This resistance will be felt by the variables that are present on your shop floor and in the environment of your facility. After a few months of your system becoming operational, I would recommend performing some

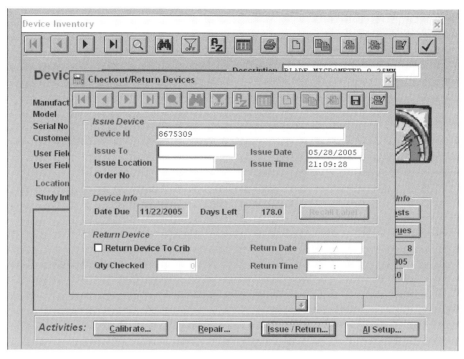

Figure 14-7: Shortcut to issue and countdown to next calibration due date. Image courtesy of GMS for Windows by Prompt Consultants, Inc.

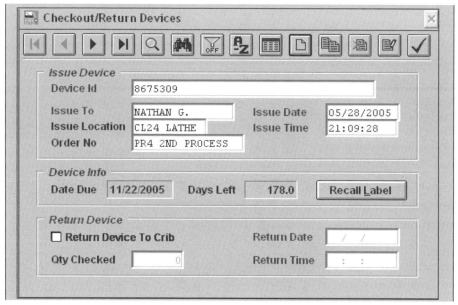

Figure 14-8: Saving the issue record after calibration and issue. Image courtesy of GMS for Windows by Prompt Consultants, Inc.

Chapter 14: Developing a Smooth Flow 131

intense internal auditing. No system will ever be perfect, but the next best thing is knowing that no one, and I mean no one (auditor, customer, member of management, etc.), will ever discover any flaw in a system that you are not already aware of and attempting to improve.

I suggest walking out on the shop floor and randomly selecting 25–50 instruments and recording their calibration status, as displayed on the floor, their specific location, and their calibration due dates. After this list is completed, you can return to your PC and confirm that the information in the software database "matches" the actual shop floor situations. Do not lose sight of the standard of precision measurement—no mistakes are acceptable. If you find one instrument out of 25 that grew legs and walked out on the shop floor from an out-of-service drawer and was being used to confirm product quality, you have a 4% failure rate. That's always the bottom line. If you find a single error out of 50 instruments, that's not twice as good but rather half as bad: a 2% failure. This 2% (one instrument) could be used to confirm thousands of products in a month, let alone three- or six-months.

Another technique is to generate a full-blown list per gage type, and perform a top to bottom audit of every instrument on the floor every three-months or so (*Figure 14-9*). This process is priceless and can teach you what your system is doing and how your software is managing it.

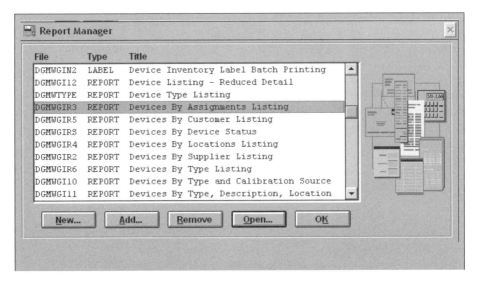

Figure 14-9: An aggressive report(s) menu allows more effective audits of the system. Image courtesy of GMS for Windows by Prompt Consultants, Inc.

This process will no doubt have critics within your facility. Those who think instruments are issued and then never move from one area to another without proper communication are living in a shell. If you have multiple instruments, multiple applications, multiple processes, and more than a handful of operators, instruments invariably will start moving in all directions as soon as they're issued. It's just like pulling your car out of the car wash… it begins getting dirty on the ride home. A quarterly top-to-bottom audit will turn over all the cards, lift up all the rugs that items have been swept under, and expose many of the weaknesses inherent to systems that involve so much human interaction.

As each weakness is discovered, I would suggest putting actions into place to countermeasure it. It is your choice if you make these countermeasures "documented," as in added to your system for the record. I like the approach of trying different variations of the countermeasures before adding controlled versions of them into your procedures. This approach has several benefits. At the top of the list are two very helpful ones. First of all, a little trial and error always produces a better process. The second may be considered mischievous. Human errors are always the most difficult ones to countermeasure, because the minute you think a system is idiot proof, along will come a bigger idiot. If you document all your countermeasure ideas "for the record" and the weakness rears its head in front of someone else (such as an auditor or customer), you may have played your best cards already. By keeping some of the countermeasures uncontrolled until you really need them, you hold a trump card for the final trick. It's not as mischievous as it is strategic and clever. It's like music. You can listen to a guitar solo and think of it either as a 30-second piece of music or the end result of 20-years of practice and training.

Just remember that you can't rush art, and continuous improvement never really ends.

Chapter 15
Out-of-Calibration Situations

"Day after day, day after day, we stuck no breath nor motion, as idle as a painted ship upon a painted ocean…"—From The Rime of the Ancient Mariner by Samuel Taylor Coleridge

Day after day, facilities everywhere are creating idle systems that paint themselves into a corner when an instrument is found not to be within accuracy specifications. It's about as easy to get people to discuss the concept of instruments found outside of established specifications within their facility as it is to get someone to talk about what is really in a hot dog while they are eating one at a cookout. Unless your accuracy tolerances are ±.500" (or ± 12.5mm) two things are certain;

1) Eventually, out of calibration accuracy situations will occur.

2) It does not have to be the end of the world or the demise of your system.

A chain is only as strong as its weakest link, and, when out-of-calibration situations occur, your system is only as good as it's ability to control the integrity of your instruments and quality of your products. We can begin to prepare by looking at established quality practices when other situations occur.

A good place to begin would be the difference between Quality *Control* and the sometimes less common term, Quality *Assurance*. Quality Control is what we practice when defective or suspect products are produced and we use different procedures in a "fire-fighting" mode to keep the defects from being processed any further. Quality Assurance is when we have systems in place that act as "fire-prevention" tools to keep the suspect or defective products from ever being produced. One of the common Quality Control measures is to evaluate the unique situation at hand and use a Quality Assurance approach to make sure the correct decisions and/or revisions are made. The actual parts, actual situations, and actual causes/effects are evaluated.

Parallel to this approach, think in terms of *control* of your measurement instruments by keeping them within a calibration interval window, and *assurance* that the instruments are currently within calibration accuracy limits. Initially, this may sound like we're talking in circles, but in this chapter you'll develop a good understanding of why this approach makes good sense.

To discuss both concepts within out-of-calibration situations, we'll use some of my favorite experiences dealing with the control of precision measurement equipment. One of the craziest examples involves granite measurement plates that are located on almost every shop floor in the world. These plates are made and calibrated in several accuracy grades: Grade AA is commonly called Laboratory Grade; Grade A is commonly called Inspection Grade, and Grade B, which is usually referred to as Toolroom Grade. I highly recommend that when these are calibrated on-site at your facility, you pick the technician's brain to learn about how these plates are measured and documented using a Moody plot.

If we look at the overall flatness differences between the grades, you'll realize that 99.99% of the time in an industrial environment you'll never be using instruments (on these plates) that can even detect the flatness difference between a Grade AA and a Grade B plate of the same size. Read the last sentence, as many times as you need to until you understand it 100%. Many times, I've seen facilities order new measuring equipment without comprehending this fact. In the purchasing process they imagine they'll get twice the accuracy at a fractional cost increase, and their business sense takes over and jumps on the higher accuracy choice. When the plates are set up in the shop, everything seems fine until the annual recalibration when, because of normal use, the calibration technician informs you that five of your 10 granite plates on the floor have worn out of their stated accuracy. The plates were purchased as Grade A or Grade AA plates, and now they're "out of tolerance."

Another common case occurs when a facility decides to buy a very nice, top of the line, set of ceramic gage blocks to use as their "grandfather set" in which most or all in-house measurements will be traced to. Gage blocks are available in several different accuracy grades or classes. It seems like a wise choice to buy a more accurate set when the set of the lower class costs between $3500 and $4000, and a set two- or three-classes "better" is only a fraction more. Once again, I suggest looking at the accuracy classes and plugging the numbers into some real world equations. If you look at a 25mm–50mm gage block along with the GGG-G-15c Gage block accuracy standard (*Figure 15-1*), you'll notice the Metric gage block tolerances are listed in micrometers. 1 micrometer = 0.001mm (39.37 millionths of an inch). A 25mm Grade 0.5 gage block must be within ± 0.04 micrometers (or ± 0.00004mm); while a 25mm Grade 3 gage block must be within plus 0.3 micrometers (or + 0.0003mm) and minus 0.15 micrometers (or − 0.00015mm). This means the difference between the Grade

0.5 block and the Grade 3 block is a whopping 0.19 micrometers (or 0.00019mm)! I don't know what you'll ever be measuring in an industrial lab where less than two-tenths of a micron will even come close to being noticed. I can say this with 100% certainty because if your 25mm Grade 0.5 gage block increased in temperature from 68.0° F to 69.3°F (or 20.0°C to 20.7°), it just expanded to over 25.00019mm. Remember Chapter 7:

Change in Length = Original Length × Change from 20° × CTE or,

25.00000 × 0.7°C (Change from 20.0°C) × 0.0000115 (CTE for Steel) = 0.000201mm

Expanded Length = Original L + Thermal Expansion or,

25.000000 + 0.000201 = 25.000201mm

In fact, if I were your auditor and you held your chin up and told me that all of your calibrations traced back to Grade 0.5 Gage blocks, I'd pull a calculator out of my pocket and explain how none of your gage blocks were within Grade 0.5 unless your thermal control of your lab was within 68° ±0.5°F.

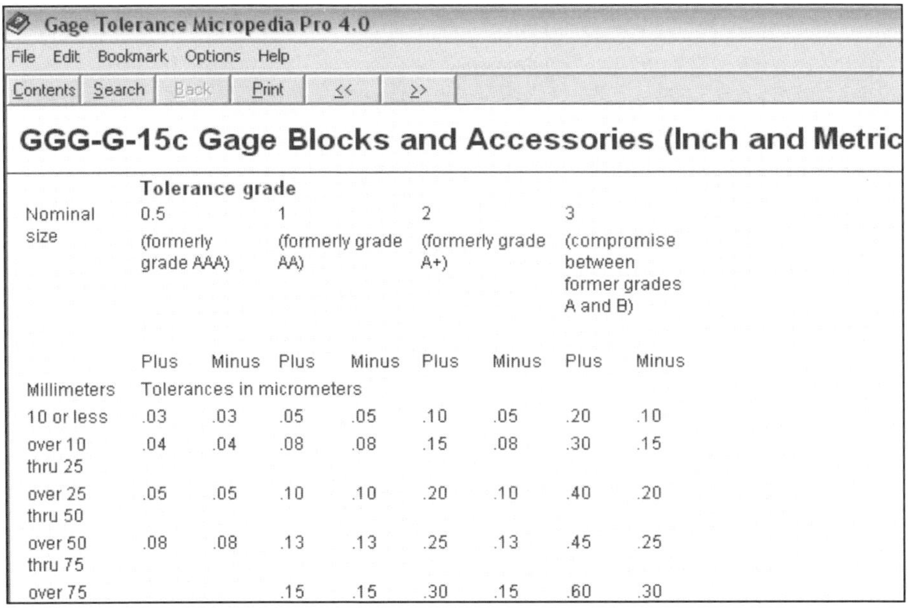

GGG-G-15c Gage Blocks and Accessories (Inch and Metric)

Nominal size	Tolerance grade							
	0.5 (formerly grade AAA)		1 (formerly grade AA)		2 (formerly grade A+)		3 (compromise between former grades A and B)	
	Plus	Minus	Plus	Minus	Plus	Minus	Plus	Minus
Millimeters	Tolerances in micrometers							
10 or less	.03	.03	.05	.05	.10	.05	.20	.10
over 10 thru 25	.04	.04	.08	.08	.15	.08	.30	.15
over 25 thru 50	.05	.05	.10	.10	.20	.10	.40	.20
over 50 thru 75	.08	.08	.13	.13	.25	.13	.45	.25
over 75			.15	.15	.30	.15	.60	.30

Figure 15-1: Gage blocks accruacy grades. Image courtesy of GMS for Windows by Prompt Consultants, Inc.

Do you see the logic? In order to get the better accuracy you paid $500 for, you'd better have spent a million dollars on the environmental control system in your inspection lab!

You may be asking right now what this has to do with "out-of-calibration" situations. Have you ever thought about how little use you'll get out of that 25mm gage block before it wears down past the 24.999996mm (–0.04 micrometers, or – 0.00004mm) lower limit? I'd guess very little. So, for a few hundred more dollars you attain a gage block set that's "twice as accurate… what a deal." This set of blocks will arrive in a very handsome, lacquered finished, felt lined, wooden box that may even have an attractive bronze plaque on the front of the box that reads: "104 pc. Metric Gage Block set: GGG-G-15c Grade 0.5." Now this won't look near as nice when the set is recalibrated in 12 months only to find that 25% to 40% of the blocks are no longer within Grade 0.5 specifications. My suggestion to you about gage blocks would be very similar to my suggestion about granite plates: Do your math homework to find out how accurate you really need to be and purchase the lowest grade possible to meet or exceed this accuracy, period! You'll have much less explaining to do later. If by chance the purchasing decision is made without your input, and your facility lines up one of these great deals where you get "twice the accuracy for a few dollars more," don't panic. In this case, do everything you can to make sure your internal documentation and protocol standards (discussed in Chapter 3) specify the lower grades as your documented internal accuracy standards. When the plates and gage blocks are outside (out-of-calibration) the higher grades during recalibration, you can just make a very simple notation on the certifications pointing out they are still within your internal accuracy standards. You can usually request the outside calibration lab reference the lower grade on the calibration certificates.

I think we've safely explained the prevention concept when dealing with granite plates and gage blocks. Where you're really going to get better is by implementing some simple quality assurance techniques for **preventing** out-of-calibration situations from going undetected by your precision measuring instruments used on your shop floor.

First, let's think of a worst-case scenario: An instrument is calibrated today, and somehow adjusted tomorrow in a way that voids the calibration and alters the accuracy. Then the instrument is used for 178-days before this error is detected. Here's the real catch: If you find an instrument is out of specification during calibration, you must somehow find a documented "change point" that you can use to isolate the instrument's accuracy shift. If you cannot prove (without a shadow of doubt) when the instrument shifted, you must assume the shift occurred immediately after the most previous calibration because that is the last documented time the instrument can be proven accurate.

And, if you thought the scenario could not be any worse than discovering an out-of-calibration accuracy instrument used for five-months and 29-days, now you must find a way to confirm the products this instrument was used to measure. Most of these are probably at your customer's facility, or have already been used in your customer's products. I didn't make this up, it's clearly explained in the QS-9000 standard:

Access and document the validity of previous <u>inspection and test results</u> when inspection, measuring or test equipment is found to be out of calibration.

When you really understand what this could mean, it becomes one very brutal sentence.

Now that we've covered one side of the spectrum, let me explain how to build checks within your system that guarantee this scenario will not happen. Not only this, but these checks will almost guarantee that if and when an instrument does shift outside accuracy limits, it will be detected within one working production shift (1st, 2nd, or 3rd), or, in the worst case, within one working day.

Above and beyond confirming that the calibration accuracy is still intact, a solid system of using part specific reference standards (mentioned in Chapters 2 and 8) will also solidify and assure the integrity of your shop floor measurements. This will also contribute to the process of lowering amounts of operator-to-operator variation in many of your measurements. Their true mathematical magic will now be revealed.

In a production situation it's very common to have granite measuring plates located on the shop floor with several instruments also issued to the specific locations that are used to measure common characteristics as described in the customer's approved control plan. The characteristics and their specifications are listed on part specific check sheets and work instructions (however titled).

Using one sample part, you can create a reference standard that can be used along with all of these instruments. This can be clearly labeled as a controlled reference standard that stays at the process as a dedicated artifact. I would recommend sample measurements of each and every characteristic be performed by operators on all shifts. The time span of 10- to 15-days would be a good start. To eliminate part variation from these study results, I would set up the measurements to be performed at fixed points on each characteristic. After this data is compiled you can crunch the numbers to determine the one- two-, and three-sigma values of each characteristic along with the overall average.

Now that you have this data at your disposal, I would suggest printing a copy of each histogram and placing it at the production location. When each operator begins their shift, require them to measure every characteristic on the sample part and record their readings on the check sheet as a start up confirmation. After a while, everybody will get used to this process and become very confident with it.

For example: Check #5 is a distance between a datum face of the part and a groove (measured using a digital height gage and test indicator), and this characteristic measured between 5.251" and 5.253" for 52 days in a row. If the operator measures the part on the 53rd day and obtains a value of 5.247", he or she will raise the red flag <u>before they run an actual piece!</u>

I've been on the other end of the spectrum and it's pure confusion. An operator brings a gage in because "it's just not right," yet everything in the equation (the parts, the instrument, and the operators) are variables. The operator has lost confidence in the instrument based on measurements of different parts performed on different days. If the instrument performs well on a recalibration or spot check using gage blocks (nearly perfectly flat, parallel, and square artifacts), the calibration technician must either replace the instrument that measures within specifications, or determine there's not sufficient data to warrant a change. **Either way somebody feels let down by the system. If the reference standard is used correctly with a histogram and data, everybody has the statistical tools needed to communicate on the exact same page.** When an instrument is questioned, everything in the equation is now a constant. When the operator brings the gage in to be replaced, all they have to do is bring the reference standard along and, with a few follow-up measurements, the root cause can be isolated and counter-measured (by replacing or adjusting the instrument). The time and effort you put into establishing the reference standards and histograms pays for itself because you have a very narrow suspect range of parts (if any), as opposed to five-months and 29-days. It's poetry in motion.

Sometimes, if you continually strive for perfection, you'll be rewarded by a glimpse of it.

Chapter 16

The Gage is Calibrated and Issued, But the Readings Don't Seem to be Right

"Houston, we have a problem…" — Jim Lovell, Apollo 13

In a production driven environment, many sources of variation occur within a measurement. Unfortunately, this is usually interpreted as instrument variation (instead of part variation, operator variation, etc.). Because of this misperception, conclusions can be drawn and improvements suggested that can actually magnify the variation as opposed to decreasing it.

One of the common scenarios I've seen occurs when a part characteristic has a not-so-generous tolerance. Because of factors within the measurement system, such as but not limited to the one-increment rule, parts falling near the tolerance limits and judged by an operator to be within specifications are often rejected at a later process because they are judged as "near" the tolerance limit but still outside of the specification. Someone usually says *"can't we just change* to a higher resolution instrument, because it should give a more accurate reading, right?" This may cause several more problems than it solves if the root-cause of the misclassification is variation within the measurement. Precision measurement instruments are magnification devises. What happens when a variable surface that is not smooth gets magnified more? You see more variation within the surface, not less. A higher magnifier is never going to make the surface smoother. If a part has a characteristic that is difficult to control, changing magnifications doesn't make the part tolerance larger.

Higher resolution can be used, but not as a <u>complete</u> instrument change. I would recommend keeping the lower resolution instrument but adding a finer resolution instrument to help complement the measurement system. The best example of this scenario uses a Metric measurement and also exposes why, in many ways, Metric measurements can be inferior to decimal inch instruments on the shop floor. The problem occurs when the decimal is moved one place to the left because the "steps" are too great (hang with me for a second, it will make common sense).

Let's imagine for a second that we have a shop floor application with a customer part specification of 50.5mm +0.05mm, and we currently use a dedicated zero-master of 50.500mm along with a depth gage and indicator to measure the part by comparing it to the master. In the beginning, we had two common instrument choices (*Figure 16-1*):

1) Using a 0.01mm indicator, which would give us a 5:1 (tolerance to resolution) ratio.

2) Using a 0.001mm indicator, which provides a 50:1 (tolerance to resolution) ratio.

After weighing out the pros and cons, we decided to would use the 0.01mm indicator.

If we change the instrument to a resolution of 0.001mm, we place the operators in a situation where the magnification sees variations that are hidden in the lower (0.01mm) resolution. An indicator with a 0.01mm resolution never seems to lose its zero-setting to the 50.500mm master, while the 0.001mm indicator never seems to hold the zero-setting. By using an instrument that requires a zero-adjustment to be performed more often, we are guaranteed to add another form of variation within the measurement.

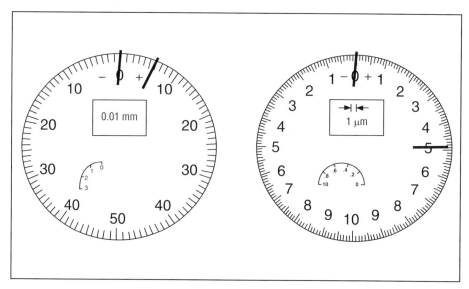

Figure 16-1: 5:1 or 50:1 ratio. Image courtesy of Mitutoyo America.

The most strategic move might be to use both instruments in a manner that maximizes speed and accuracy, and keeps confidence within the measurement results much higher. If we have both the 0.01mm and 0.001mm instruments at the process, we can use the 0.01mm to confirm each part. If the result is within specification (inside the limits) by more than one increment (50.52mm – 50.54mm), I'd accept the result and "pass" the part as within part specifications. If the part measured within one increment of the specification limits (50.51mm or 50.54mm) I would zero the 0.001mm indicator and use this instrument to make the final classification (*Figure 16-2*). This may seem (to the "can't we just" crowd) to be additional steps that are not needed, but in reality it is accepting the strengths and weaknesses of each "chess piece" and using them in the most strategic manner possible. These are not additional steps. An additional step is sorting parts because the flaws in your measurement system may have misclassified them.

Another common situation on the shop floor occurs when measuring with a standard two-point tipping (or sweeping) type bore gage. When a tight customer part specification is one of the six elements within the measurement system, a very common instrument choice is to use a Class XX or Class X ring gage as a zero-master, along with an indicator of a 0.001mm resolution (if a Metric reading) or .0001" resolution (if an inch reading). Many operators and members of management don't fully comprehend some of the mechanical flaws that are the nature of the beast within this application. For starters, when zero-setting the bore gage or measuring the part, hysterisis error will be present at these high resolutions. **Hysterisis is an error phenomenon caused when an instrument "reverses" direction.** As a bore gage is placed

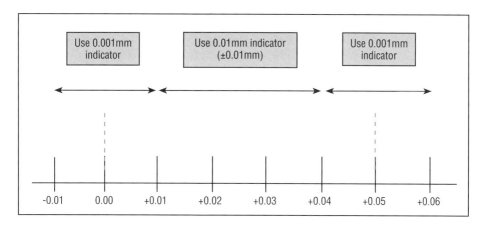

Figure 16-2: Knowing the one-increment rule can allow for better instrument choices.

inside a mastering ring, with the gage at a tilt, the contacts are extended. As the operator "sweeps" the bore gage *forward* through the ring gage, the contacts move. This is seen as a decreasing reading obtained on the indicator.

When the contacts of the bore gage are parallel with the perfect (level) diameter of the ring gage, the indicator "buries" at its lowest reading, and as the gage is moved farther through the sweeping action the contacts extend again and the reading increases again. If the sweeping direction of the bore gage is reversed and the operator sweeps the gage *backward*, the contacts begin to contract again, but when the indicator "buries" going the other direction, the reading will differ from when the indicator buries during the forward sweep. This is hysterisis. As precise as these instruments are, there is still a very small amount of "play" between the moving parts, and when a movement direction is reversed the play has to work itself out, causing a hesitation that changes the reading.

Another common problem occurs when some operators choose not to use the sweeping techniques, and instead just stand or balance the bore gage at what they feel is a perpendicular position. This technique is flawed for at least two reasons. First, even if the bottom face of a tipping bore gage is **perfectly** perpendicular to the bore gage stem at purchase, it won't be after a few weeks of use on the shop floor because the bottom of the bore gage will begin to wear and develop burrs and other flatness variations. Secondly, no two operators in the world will have an identical "feel" for which position is perfectly perpendicular. Any amount of pitch, yaw, or roll will cause

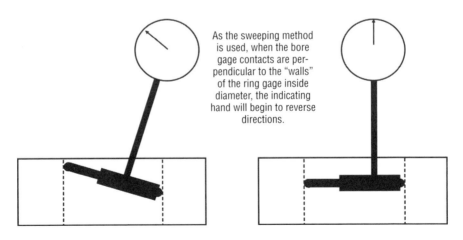

Figure 16-3: When finding the smallest diameter of an inside bore, the sweeping technique is recommended.

the measurement results to vary. When this occurs, borderline parts will become misclassified and the integrity of the gage itself will be questioned.

To correct this situation, we must follow the gage maker's recommendations and use a sweeping technique in our measurement method. To account for the hysterisis is very simple. We must decide if we are going to reference our zero-setting and our part measurement on the forward sweep or the backward sweep. As long as you measure the same as you zero-set, the hysterisis is taken out of the equation. Once this direction is decided, we must conform to it. I've seen many times when people want to have their cake and eat it too in the form of using the hysterisis error of the other sweeping direction when it makes a borderline (but out-of-tolerance part) become a borderline (but in-tolerance) part.

If these improvements don't get your facility where it needs to be with a specific measurement application, I would consider an application-specific instrument. So much of developing better measurement systems involves identifying constants within the system (and keeping them controlled), while at the same time identifying the variables (or changes) and trying to find ways to control them better. If we were to experience the bore gage example we just considered, and we could not keep up with the variables, we could consider a non-tipping bore gage (I have had very good experience with these products, and there are several competent gage makers who can provide application-specific instruments). A non-tipping bore gage (see *Figure 16-4*) does not require a sweeping action to measure a part diameter, and the gage shown in the Figure comes with a less than 10% GRR guarantee.

This guarantee is a concept that we should take some time to examine. The manufacturer of the gage in the Figure (Dyer Gage) sends along data documenting that the gage will posses under 10% GRR variation when measuring dedicated masters within your part's tolerance. This capability doesn't seem to fit within the six-elements of a measurement concept we discussed in Chapter 7 (because masters are used instead of sample parts), but it is not intended to. What this documentation proves is the repeatability of the gage: it will *consistently* perform within a less than 10% GRR level. My father-in-law once told me there are two types of people in the workplace: assets and liabilities. And, if someone is 49% asset, they're still a liability. In measurement, there are two types of factors: constants and changes (or variables). Anytime you can improve a variable to make it 51% constant, you have tipped the odds closer to your favor. Instruments, like Dyer's non-tipping bore gage, make the measuring process as close to a constant as you can get. If the instrument becomes a constant within the measurement, you can focus and hopefully gain better control of

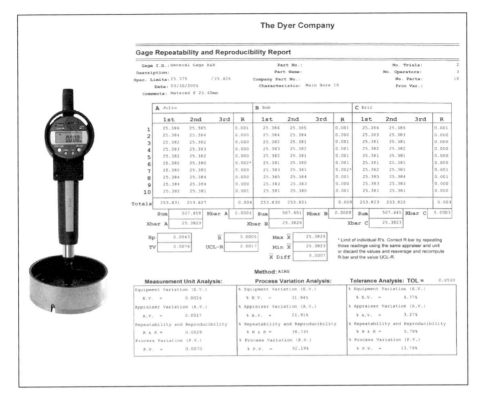

Figure 16-4: 10% gage R & R is guaranteed. Image courtesy of The Dyer Gage Company.

the other changes (or variables) within the remaining five-elements of the measurement. When this is complete you'll know the instrument is very capable of obtaining a less than 10% GRR.

This philosophy demonstrates the most common misunderstanding about calibrated instruments and measurements performed with the instrument on the shop floor. Calibration is a controlled test in which five- of the six-elements within a measurement system are taken out of the equation in order to determine a "best the gage will perform" level. This level of accuracy becomes the constant, and in order to obtain anywhere close to this accuracy within your shop floor measurements you must control the changes within the other five-elements.

I've observed this misunderstanding when operators are using height gages at side-by-side locations. Digital height gages have an accuracy specification of about 0.025mm (or .001"). This becomes very prevalent after a few years of daily use on the shop floor. These gages sometimes seem to develop a personality of consistently measuring either high or low. If operators are measuring similar parts at different locations, it can become rather chaotic. If a part has a machined step that calls for a distance of 61.6mm +0.1mm, the error between two height gages can consume more of the tolerance than most are willing to accept. Operator A will measure an entire layer of parts that read 61.68mm or 61.69mm. The next day, when operator B attempts to run the parts through the next process, he measures them to be 61.72mm and 61.73mm (out-of-tolerance). Frustrated operator A pages the calibration technician to the shop-floor and demands to know "which gage is right?" Then he becomes really frustrated when the calibration tech measures a 100mm gage block at 99.98mm with gage A and 100.02mm with gage B and says "they both are."

I've had Quality Managers argue this point. Once, I even had a manager insist that I show him this variation because he was "pretty sure" the O.E.M. accuracy specifications were the *error allowed at the maximum measurement range*, as opposed to anywhere within the range, which in this particular case would have been 300mm. (Typical accuracy ranges, in this case for a digital Mitutoyo height gage, are shown in *Figure 16-5*) To prove his point, here is how he insisted we perform the test: First, he wanted the two height gages used in the quality lab, which happened to be the newest height gages in the plant, zero set to a 100mm gage block. Next, a 25mm gage block was wrung on to the 100mm block, and each height gage was used to measure the 125mm gage block stack. Finally, he had us "practice" until we could "make" each gage read 25.00mm—if we measured 24.98mm, 24.99mm, 25.01mm, or 25.02mm, he

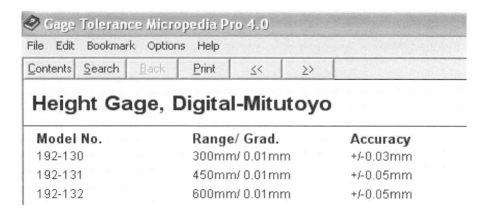

Figure 16-5: Height gage accuracy listed by OEM. Image courtesy of Gage Tolerance Micropedia by Prompt Consultants, inc.

would say, "well, maybe you bumped it or something, measure it again." So, after we "created" a 25.00mm measurement with two brand new height gages when measuring gage blocks, he informed me this proved the gages were more accurate than the listed O.E.M. specification. So he proved my point and didn't even realize it: How an instrument measures a near perfectly flat, parallel, block of steel (of a known value) in a controlled laboratory environment is not a good benchmark to predict how the instrument(s) will measure a production part of unknown value on the shop floor.

What's really ironic about this scenario is that we both understand it, and we accept a small degree of uncertainty in every other situation in our lives. I've never seen an agenda for a meeting that is scheduled to start at 9:02AM, nor have I found the hours of operation for a business that claims to open promptly at 7:57AM. This is because we will always argue about whose watch is correct at 9:02AM, or what clock to go by at 7:57AM. If we have a very important deposit to make on a specific day at our bank, we don't intentionally arrive there when our watch says 4:58PM, and then argue with security about what time it is when he's says "sorry, we closed at 5:00PM." We accept uncertainty in what time of day it actually is, and we err on the side of caution and get to the bank 10- to 15-minutes before closing time just to be safe. Yet every day on shop floors all over the world, operators are running processes right on the razor's edge of their allowable tolerances and starting arguments about increments that are fractions of a fraction of the thickness of a human hair. This is the reality that must be faced: Because of error, variation, and uncertainty within any and every measurement, we cannot realistically expect to use 100% of any tolerance. If we do, we'll always be chasing fractions of a millimeter or ten-thousandths of an inch, and winding up exactly where we started: wondering what our parts **really** measure and hoping they are within specifications.

Like it or not, it is what it is.

Chapter 17
Evaluating Training Needs and Concepts

What you hear, you forget… what you see, you remember… what you do, you understand

Even the best measurement control system in the world will not improve the quality of the products that are measured and confirmed unless the operators, employees, or associates (however titled) are trained effectively and evaluated correctly. Too often, in some facilities we think that if someone keeps falling off of a bicycle, we must place training wheels on the bike. If they still fall off, we put training wheels on the training wheels. If they still can't stay on the bike, we ask people to jog along-side them to keep them balanced.

Precision measurement is often as much of an art as it is a science, and a good measurer will often posses a fair amount of finesse when performing his or her measurements. In this chapter we will discuss three topics:

1. Evaluating measurement ability,

2. Selecting the best candidate(s) for advanced training, and

3. Establishing your facility's shop floor measurement philosophies.

It is best to discuss training or evaluation scenarios that may (and most likely will) occur at your facility, and how <u>not</u> to handle them. If you can understand how not to handle them, the most effective way of handling each situation will become more obvious. These situations are where the chess approach can be so vital. Each possible move should be thought out in a cause-effect manner. We can strategically predict the best and worst case scenarios for any move, and decide on which one has the highest probability of success and the lowest probability of failure. Just like chess, if you gamble on too many bad moves the game may soon domino into a no-win situation. These examples are best explained through actual situations that I have been fortunate enough to experience in the past few years. Only the names have been changed to protect the guilty.

Our first example demonstrates how a perception, no matter how widely accepted, cannot turn day into night or night into day. One of these perceptions is that if someone works at a trade long enough, they are automatically skilled and most likely an expert. This is made possible by an incorrect belief that using precision hand measuring tools **correctly** is a task simple enough for your average monkey to perform.

Several years ago, I was working as a consultant implementing an ISO-9000 compliant inspection, measuring, and test equipment system in a small (about 20-machinists) family owned tooling facility. Very early in the calibration process, I began calibrating a personally owned set of micrometers. The set contained six friction thimble, inch reading micrometers, .0001" resolution. When I zero-set the 1" micrometer, I noticed it read minus .0001" which I attributed to a difference in "feel" between the operator and myself. I then took the required measurements throughout the 1" range and found the micrometer to be very accurate. When I zero-set the 2" micrometer to a 1" gage block, I noticed it read plus .0002", which didn't make sense. If the operator and I had a different feel, as I thought was the case with the 1" micrometer, it should be (but was not) consistent. The 2" micrometer calibrated very well also. As I calibrated the rest of the set, I discovered that each micrometer had a different deviation at its zero-setting. Anyone who calibrates hand tools six- to eight-hours a day using gage blocks—which are the flattest, squarest, and most parallel artifacts in the shop—usually gets very consistent with measuring these highly accurate pieces of steel or ceramic.

In this situation I was presented with 2 possible moves:

1. Leave the micrometers set at inconsistent zeros (which guaranteed inconsistent measurement results), or

2. Create a cardinal-sin by adjusting someone else's micrometer to my zero, which I knew to be consistent.

Isaac Newton once stated that few people really understood that when choosing the lesser of two evils, they are still choosing evil. I adjusted the six micrometers to my zero-settings and explained this to the operator when I returned the instruments. As I calibrated micrometers from other operators I discovered the same inconsistencies.

Later, back at the shop, the Vice-President was holding a meeting with operators to discuss unacceptable occurrences of customer products being returned because of out-of-tolerance features. The operators were quick to suggest that the root cause might be due to the "calibration guy" adjusting their zero-settings, which could cause incorrect measurements. I received a phone call from the (understandably irate) VP later that evening. He went into great detail about how skilled his operators were and how dependent measurements were to an operators "feel" for his or her gages. This is the perception that most people have: measuring is simple, and everybody can do it. Furthermore, since his operators were veterans, they were consistent and skilled measurers. I asked the VP if I could have 30-minutes of time split between two-days to investigate this situation and we could discuss the facts and proceed from there.

The next morning I found a Class XX plug gage and compared it to gage blocks using a very accurate bench micrometer. The plug gage was an odd size of 1.4562". Even though a Class XX plug gage is extremely round and cylindrical, I marked a fixed-point diameter across the plug to eliminate any "perceived" part variation, as shown in *Figure 17-1*.

I then walked out on the shop floor with a 1" ceramic gage block and the plug gage placed on a shop-cloth to decrease thermal effects. I separately asked 10 operators to zero-set their micrometers to the 1" ceramic block and measure the plug gage across the fixed point diameter. After collecting this data, I charted a histogram (*Figure 17-2*) documenting the readings (average and range). The range of values obtained between the 10-operators consumed .0005" (5 tenths).

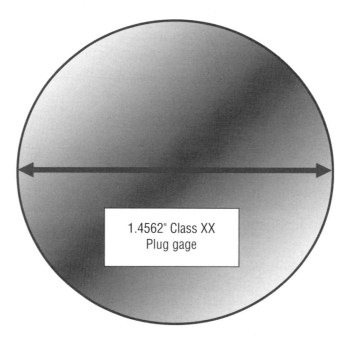

Figure 17-1: Plug gage measured at a fixed-point diameter.

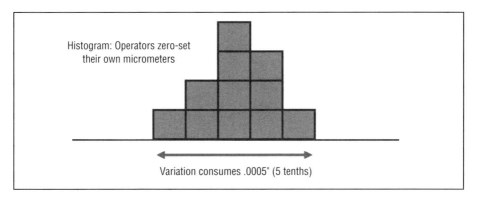

Figure 17-2: Histogram of measurement data when operators zero-set their own micrometers.

The next morning, I found another Class XX plug gage and compared it to gage blocks using a bench micrometer. This plug gage was also an odd size, of 1.4581". I marked a fixed point diameter on this plug gage as well. My next step was to find a very low market, inexpensive, mass-produced two-inch micrometer and zero-set it **to my feel** in the lab. Then I went out on the shop floor and handed the cheap micrometer and plug gage to the same 10 operators and asked them to measure it across the fixed point diameter, and I charted another histogram (*Figure 17-3*) documenting the readings (average and range). The range values obtained between the 10-operators consumed .0003" (3 tenths). *That's right!* **The operators actually experienced 40% less variation when measuring with a micrometer that was zero-set for them.**

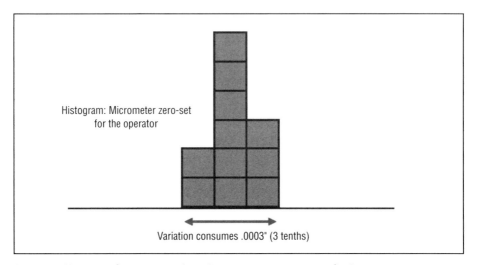

Figure 17-2: Histogram of measurement data when micrometer was zero-set for the operator.

152   Exposing the Myths of Industrial Precision Measurement Control

I met with the VP and explained to him that he indeed had a problem. His operators, whom he felt had the "feel" of an expert, apparently *could not effectively zero-set their own instruments*!

The solution to this problem was very simple. We obtained steel gage blocks to use as setting masters (1", 2", 3", etc.). These gage blocks were calibrated and stored in a central location on the shop floor. We also machined and ground some round reference samples (Ø.7891", Ø1.4185", Ø2.3526", Ø3.1564", etc.) to use with the gage blocks. These could be any size within the range of each micrometer, as long as the diameter value was established. The reference samples were measured in the inspection laboratory using a bench micrometer. After the operators zero-set their micrometers to the steel gage blocks that were designated as setting masters, they could measure the reference sample to confirm the result obtained today was comparable to what was obtained yesterday, the day before, or last week. We could "test" anyone's ability to measure the reference samples against the accepted value. This method lets everyone understand how they measure when "grading on the curve," which was the reference value and the average obtained by all the operators in the shop. If people have a tendency to measure low or high they can account for this and truly develop a consistent and repeatable technique.

It is important to understand that these things have always been hidden in your measurements, and you shouldn't be surprised when you discover them. A common parallel can be found when a production shop starts an aggressive scrap reduction plan. Usually the first thing that happens is scrap amounts increase after the plan gets put into place. Why? Because the scrap they've always had is getting documented correctly now. People who think the scrap amounts really went up are blind. The truth is the scrap amounts were probably always as high, but it was getting tossed out back in the hopper while nobody was looking instead of being accounted for on the bottom line. You're always going to discover things you've never noticed when you look in areas you've never looked.

Our next example of how not to approach training involves off-site training classes or seminars. I have a very close friend; we'll call him Peter Lemmonjello, who is very adept with coordinate measuring machines (CMMs). A few years ago, the facility where he works full-time was up-grading from a manual CMM to a top-notch DCC CMM (CNC Programmable). The machine uses a trigonometry based DMIS code in the part programs (ppgs). My friend had more CMM experience and knowledge than everybody else in the 250-employee shop combined. When it came down to selecting the three-individuals to attend the five-day training class given by the OEM, his management decided on who they felt would be using the CMM the most and sent them. He explained to the "powers that be" that it would be better for the facility to send the best candidate available to absorb as much knowledge as possible and bring the knowledge back to be taught to other individuals when those individuals were ready. Management, who *knew* about 2% of what they thought they knew about coordinate measuring, would not let go of their ability to control the outcome and "decide" on **"*their* choices"** to attend the seminar and return to the facility "*trained*." Needless to say, the three-individuals came back with their brains melted. The advanced CMM concepts of the DCC CMM and its software were way above their experience level, which consisted mostly of measuring three-point circles. This is the industrial parallel of management "deciding" they can choose "their" painter to repair the cracks in the wall, and, because of some divine power, the cracks will magically not reappear because they're convinced the shifting foundation is nothing but a fable (this won't make sense if you haven't read the Introduction).

It's been five-years since this experience, and my friend has written several magazine articles about DMIS CMM programming and is currently in the process of writing a book. No matter how much the deck was stacked against him, he was not going to be denied. Any good off-site training seminar should cram the attendees full of information that can slowly be discovered as the experiences bring out the knowledge later. If you're in management and you really think you can decide who learns and who doesn't, you may be mistaken. I'm a believer that 99% of accomplishing any goal is wanting to. Those who want it bad enough will rise above the rest, whether they're the "chosen" one or not. If you really want to make the smart training choice, let the enthusiasm of your employees make the decision for you.

Bill Parcells, the architect of the 1986 and 1990 Super-Bowl champion NY Giants, once said when his team would get close to the goal-line with the game in the balance, he would never think about <u>plays</u>, he'd think about <u>players</u>. When all the marbles were on the line, he thought of which players had established themselves as who you'd put your money on to come through in the clutch. This is how to get the most bang for your buck out of your (sometimes very costly) outside training opportunities. Think of who has established the "fire in their eyes" and let them know you are counting on them to go to the training and bring it back to train the masses.

This concept has a hidden dual benefit. Several years ago I was fortunate enough to spend four years of my life in the 82nd Airborne Division. I knew the individuals in my chain-of-command were good, but I realize now they were among the best leaders in the world in my generation. One concept always stressed was <u>training the</u>

trainer. Their philosophy was: if you thought you had something down to a science, go teach it. **You learn more by teaching than you'll ever learn by being taught.** Send the individuals to training who have established themselves as those who want the knowledge and experience the most. All you have to do after they get back is stay out of their way and give them the opportunity to work their magic. Trust me, if you try this approach once, you'll never look back.

One final concept and example is older than I am**:** Keep it simple, stupid! This concept is often mentioned but rarely followed. Facilities think they'll make it simple and then proceed on complicating items in the long run. Let me offer this example. In a production shop, characteristics of parts are often measured and controlled using dedicated part specific masters. Indicators are zero-set (or mastered) to the master and the part characteristic is measured by comparing the deviation from the nominal value of the dedicated master. It may often seem "simple" to record all of the readings on the part characteristic check-sheet (however titled) in a deviation format as opposed to an actual reading. For example:

If the part tolerance were 79.1 ±0.05mm (*Figure 17-4*), and we measured the part after zero-setting the (0.01mm indicator) to the (79.100mm) master and our indicator read 0.02, we could record only the deviation (+0.02mm) as opposed to the actual value (79.12mm).

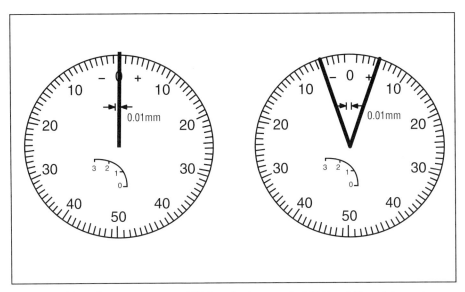

Figure 17-4: If the indicator was zero-set at 79.10mm (as shown on the left) the ±0.05mm tolerance is shown on the right. Image courtesy of Mitutoyo America.

This is simple in the short term, but has its demise in the long term. Eventually, operators lose site of what the part reads and only pay attention to deviation(s). Not long after this, they'll begin to lose sight of the resolution of the indicator and no longer see the deviation as +0.02mm, but instead it will become "plus two." Before long people forget what the part specification is because your system allows them to not **really read** the measurement. When this happens, the dominos fall in every direction. Out of necessity someone will suggest we "mark" the indicator to easily identify whether the part is within tolerance or not. This also sounds simple, but now the operator doesn't even have to read the gage at all. Instead, they simply pay attention to whether the reading is in the "green" area of the gage or the "red" area.

I don't want to beat this subject up much more, but there are some wonderful metaphors for these concepts found "in the movies" (hang with me here, it really gets the point across). A few years ago there was a movie called "Armageddon" in which a meteor roughly the size of Texas (called a global killer) is on course to impact our planet. Our only chance is to land on the meteor, use oil drilling techniques to drill into the meteor and drop a nuclear explosive device into the core of the meteor that, when detonated, will break the meteor into two pieces and send both pieces into a safe trajectory.

At one critical point in the movie, the drilling is not working; the President imposes Martial (Military) Law and takes over NASA. He thinks because they are running out of time, the device should be detonated on the surface of the meteor. The scientists at NASA know a surface detonation will not affect the meteor and at one point in the movie a very passionate NASA official says "Mr. President, if you detonate this device on the surface you will guarantee the failure of this mission."

This method of thinking is priceless. Marking gages so that operators don't have to learn and stay proficient at reading them, equates to nothing more than training wheels. All you will accomplish by keeping training wheels on a bike (or investing in more elaborate training wheels) is to guarantee that the person will *never* learn to ride the bike.

As lucky as I've been at being able to put thoughts into words in my short writing career, I could never sum this mode of thinking up as well as Scott Paton, Publisher for Quality Digest magazine. Scott recently wrote an editorial of sorts about the retail juggernaut Wal-Mart, and about how no matter how much money is invested in a customer service plan or concept, it will always be judged where the rubber meets the road:

"Sam's baby has state-of-the art software systems that manage its distribution. It has talented people who decide where to locate its stores for maximum return. It knows exactly how much of each product to order to maximize sales and minimize inventory. Yet when I walk into a Wal-Mart at midnight and I can't get my questions answered or I can't find what I'm looking for, those state-of-the-art systems are worthless."

Shop-floor instruments are simple enough to read. If you have operators that need it simplified more, consider using some of the training philosophies discussed in the next chapter, and if that doesn't work, consider interviewing some new operators.

In the long run, all the "simple" shortcuts provide more opportunities for abnormal situations to occur, yet go unrecognized because everything is dummy-proofed to the point that it has created dummies. In the next chapter we will learn how to use the "clock method" to demonstrate how simple methods of teaching proper instrument reading can be derived from the steps we all learn when we were taught how to tell the time of day.

It's just as simple as counting to 10. Don't make it any more difficult.

Chapter 18
Training "By the Numbers"

"As Long as I remember the rain's been coming down.
Clouds of mystery falling, confusion on the ground.
Good men through the ages, trying to find a sun.
And I wonder, still I wonder, who'll stop the rain?"—John Fogerty

At the age of 19, I was fortunate to visit the Egyptian pyramids on the Giza plateau. Since then, I have since seen many interesting stories on television marveling at how an ancient culture could build such geometrically precise structures. Having been there, I am more amazed at how they could even design them to such mathematical accuracy and precision. The Egyptians either solved the most difficult engineering and trigonometry scenarios in the history of mankind (by drawing in the sand) or they possessed some fundamental knowledge of numbers that we have, in our world of cellular internet, missed. I would guess both.

Let's step outside the box and look at an example of the simplicity and beauty of numbers. I don't know how I made it through 12-years of public education in this country without being taught the following, which can be found in the first 50-pages of just about any number theory book.

If we add consecutive odd numbers, the sum will always equal the next perfect square.

$1 + 3 = 4$, the square root of $4 = 2$

$1 + 3 + 5 = 9$, the square root of $9 = 3$

$1 + 3 + 5 + 7 = 16$, the square root of $16 = 4$

$1 + 3 + 5 + 7 + 9 = 25$, the square root of $25 = 5$

$1 + 3 + 5 + 7 + 9 + 11 = 36$, the square root of $36 = 6$

$1 + 3 + 5 + 7 + 9 + 11 + 13 = 49$, the square root of $49 = 7$

$1 + 3 + 5 + 7 + 9 + 11 + 13 + 15 = 64$, the square root of $64 = 8$

$1 + 3 + 5 + 7 + 9 + 11 + 13 + 15 + 17 = 81$, the square root of $81 = 9$

1 + 3 + 5 + 7 + 9 + 11 + 13 + 15 + 17 +19 = 100, the square root of 100 = 10

And it never ends!

Here's another, if we sum odd numbers in increasing intervals (first 1, then the next 2, then the next 3, then the next 4, etc.) the cubed root of the sums are in perfect order.

1 = 1, the cubed root of 1 = 1

3 + 5 = 8, the cubed root of 8 = 2

7 + 9 + 11 = 27, the cubed root of 27 = 3

13 + 15 + 17 + 19 = 64, the cubed root of 64 = 4

21 + 23 + 25 + 27 + 29 = 125, the cubed root of 125 = 5

And this one never ends either!

Now, you're probably asking questions such as "what do these examples have to do with solving a triangle or balancing an equation," and the answer is: nothing. That's probably why it's not taught, because it's only use is to get you to think in terms of simple patterns within a complex sample or system, which is often viewed as taboo or something we "don't really need to know."

We can apply this same principle of simple patterns within samples or systems to learn the most effective ways to teach how measurement instruments are read. This can be seen in the "clock method." If you can read a clock, you can teach someone to read most measurement instruments in about three-minutes, using just two-steps. Actually, there are over twice as many steps involved in learning how to read a clock as there are in learning how to teach the reading of most measurement instruments, yet most of our first-grade children can probably read clocks better than many industrial professionals "really" read instruments.

A clock is the first variable instrument we learn to read. The indicating hands on the clock are read by their current position on the scale located on the clock or dial face and, through a process of addition, the time of day is determined. If you look at the clock in *Figure 18-1*, you can most likely tell in a glance that it reads 1:32. This ability to read the instrument in a second's glance comes from years of practical application.

What we fail to recognize in these steps is a simple process that we learned at such a young age. I am not trying to insult anyone's intelligence, but to fully under-

Figure 18-1: Reading a clock is no different than reading many other variable instruments.

stand the process and apply the same process in our training methods, we should list the steps of the process "by the numbers."

• Step 1: Identify the two-increments the short indicating hand is positioned between. We must realize that when using the short indicating hand, the 1–12 scale around the dial face represents increments of one-hour.

• Step 2: Since the short indicating hand is positioned past the 1, but before the 2, we can record the hour increment of 1:00. Because the short indicating hand is past the 1, we know our time of day reading will be 1:00 "plus some change," which in this measurement will be in the form of one-minute increments.

• Step 3: Now we are moving on to reading the long indicating hand which changes the value of the 1–12 scale around the clock face. When using the long indicating hand, the scale represents increments of five-minutes. As we count around the face, one complete rotation represents a time interval of one-hour (or 60-minutes).

• Step 4: Because the long indicating hand is positioned past the 6 (which represents 30-minutes), and before the 7 (which represents 35-minutes), we can determine our reading to be 1-hour + 30-minutes + some change.

- Step 5: For our next to last step, we need to correctly identify the resolution of our instrument, which is one-minute. The long indicating hand is two-increments (two-minutes) past the 6 (which represents 30-minutes).

- Step 6: Our final step involves adding the three separate readings obtained from our 2 indicating hands when using three separate scales.

- 1 hour + 30 minutes + 2 minutes = 1:32.

We should note that we really left out an additional step of identifying whether our reading should be AM or PM.

Now, with some of the narrow minded individuals I've encountered in industry, it's a wonder of science they ever followed that many steps of instructions without changing a step and turning it into a modified step that makes sense only to them. It's the first instrument we ever learn to read, and it's without a doubt the most difficult, yet we can learn it when we are five- or six-years-old. The reason is simple: We are not old enough to be intimidated by a challenge and we don't think we know how to read it before we are taught.

The reason this represents a great instrument reading training model is because if we can just get people to take a deep breath, don't give in to being intimidated, and follow similar steps, we can teach them how to read instruments correctly. This can eliminate the need for "training wheels" or keep it simple, stupid techniques that turn counting beads on a string into applied mathematics.

To prove the validity and simplicity of this "by the numbers" method, we need look no further than some precision measurement tools most likely being used right now on your shop floor. We can teach most of these measurement tools in two-steps as opposed to six or seven. We can start with a Metric Vernier caliper having a resolution of 0.05mm. This type of instrument uses two scales: The main scale, which runs the length of the instrument—such as 150mm or 200mm—and reads in increments of 1mm, and the Vernier scale which reads in increments of 0.05mm

- Step 1: Read the main scale to determine the 1mm value. If we look at the image of our caliper (*Figure 18-2*), we can use the line above the "0" on the Vernier scale just as we use the short indicating hand on a clock. Our short indicating hand on the clock rests between the one- and two-increments, which allows us to determine the time to be between 1:00 and 2:00. The 0 line on the Vernier scale rests between the

52- and the 53-increments of the main scale. This means our caliper reading is 52mm plus some change.

- Step 2: Our final step involves following the Vernier scale from left to right until we find the increment on the Vernier scale that lines up the closest to an increment on the main scale. Each Vernier scale increment represents 0.05mm. We can determine the Vernier scale increment between the 3 and 4 lines up very close to an increment on the main scale.

- 52mm + 0.35mm = 52.35mm.

Wow! Two-steps. Why in the world do people make it any more difficult than this in an effort to *keep in simple*? We can use two very similar steps to teach someone how to read a Metric Vernier micrometer having a resolution of 0.01mm.

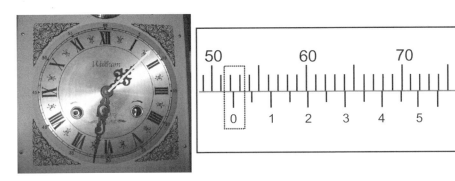

Figure 18-2: Reading the main scale of a vernier caliper is just like reading the hour on a clock.

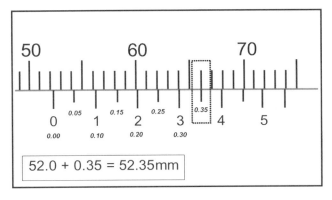

Figure 18-3: The decimal value of the Vernier scale "counts" in 0.05mm increments.

The Vernier micrometer also has two scales. The main scale, located on the barrel of the micrometer, reads in increments of 0.5mm, and the spindle, (or cylindrical) scale, located on the spindle reads in increments of 0.01mm.

- Step 1: Read the main scale to determine the 0.5mm value. If we look at the image of our micrometer (*Figure 18-4*) we can use the lines above the horizontal index line—which represent increments of 1.0mm (whole millimeters)—and the lines below the horizontal index line—which represent 0.5mm (half-millimeters). If we read these main scale increments from left to right until we run into the spindle, we can determine our 0.5mm value is 13.5mm. This means our micrometer reading is 13.5mm plus some change.

- Step 2: Our final step involves following the horizontal index line to determine where it "lands" on the spindle scale, which has a resolution of 0.01mm. In our measurement (*Figure 18-5*), our index line meets the 36-increment, which represents 0.36mm. To determine the total reading we can add 13.5mm + 0.36mm = 13.86mm.

Teaching someone to read an indicator is even simpler yet. These instruments use the clock method in its purest form. We have to stretch it to come up with two steps.

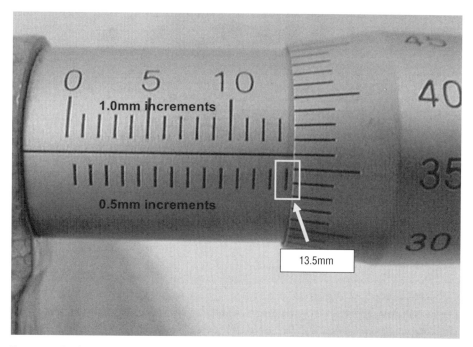

Figure 18-4: Reading the 0.5mm value of the main scale. Image courtesy of Mitutoyo America.

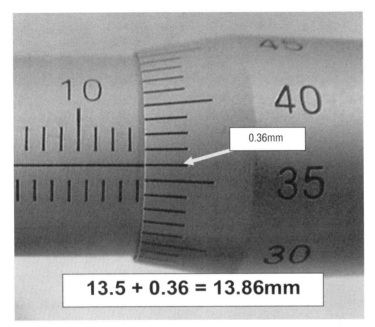

Figure 18-5: Reading the 0.01mm value of the spindle scale. Image courtesy of Mitutoyo America.

- Step 1: Identify the resolution of the indicator. This information should be located on the "face" of the indicator and will explain how much of a movement each increment represents.

- Step 2: Count the increments the indicating hand has moved, and multiple by the resolution (*Figure 18-6*).

As simple as this is, you would think that nobody could ever figure out a way to (in an effort of keeping it simple) make it more confusing or difficult, but, trust me, they do. This can occur from individuals getting used to a certain indicator configuration. Some indicators "count" in an increasing manner. As you move clockwise from the 0, the instrument count increases (0, 10, 20, 30, etc.). Some indicators count in a decreasing manner. As you move clockwise from the 0, the instrument seems to count backward (0, 90, 80, 70, etc.). Other indicators have what is termed a balanced-face. As you move clockwise and counter-clockwise from 0 the instrument count increases. Some of these differences are shown in *Figures 18-7* through *18-9*.

Many debates on the shop floor can occur by trying to "figure out" whether a clockwise movement of the indicating hand means that the part is larger or smaller. The rule of thumb is: there is no rule of thumb. You can only figure out this pattern one application at a time. Depending on how the instrument is measuring your part,

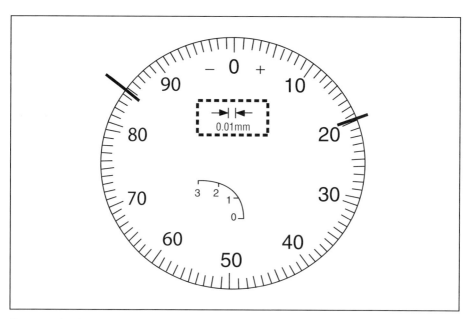

Figure 18-6: Reading an indicator: Count the increments and multuply by the resolution. Image courtesy of Mitutoyo America.

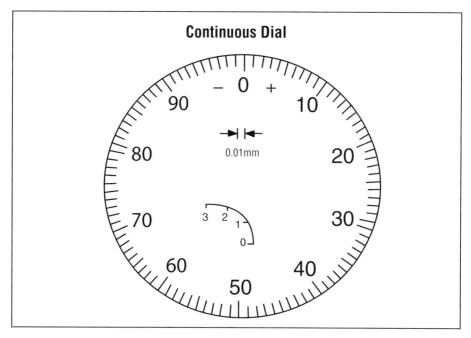

Figure 18-7: Continuous-reading indicators "count" clockwise. Image courtesy of Mitutoyo America.

and/or how the fixture and master are set up, the rotation of the indicating hand can represent an increasing or decreasing reading.

Another common problem occurs when dedicated masters are used to zero-set indicators for a part-specific application. The master is usually made to the part specification nominal, and the deviation from 0 represents where the part characteristic measures compared to the (nominal) master. What causes a problem is people get used to a certain indicator configuration. For example, if we had a zero-set an indicator to a 61.600mm master, and our part specification was 61.6mm ± 0.1 we could measure our part with a 0.01mm indicator and the reading could range anywhere from 10-increments counter-clockwise (from 0) to 10-increments clockwise (from 0), and our part would meet the specification. Notice that I stated the part could measure ± 10 increments, not *"anywhere between the 90 and the 10!"* If we teach our operators to think in terms of INDICATOR RESOLUTION and INCREMENTS, it will never matter what indicator (face) configuration they use. Many times, I've seen an operator taught that the above part specification ranged anywhere from the 90 to the 10 because at the time an indicator was being used that had a clockwise increasing count (*Figure 18-10*). When we changed part types and used a balanced-indicator on our 55.5mm

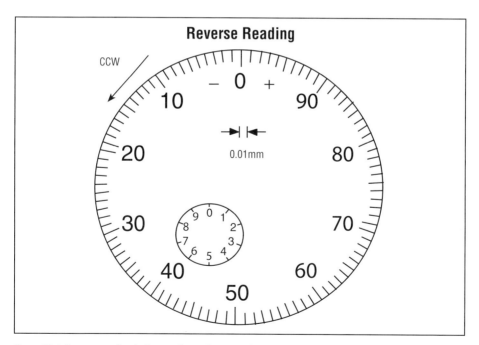

Figure 18-8: Reverse-reading indicators "count" counter-clockwise. Image courtesy of Mitutoyo America.

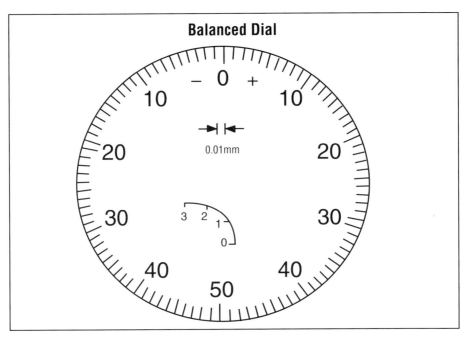

Figure 18-9: Balanced-indicators "count" CW and CCW. Image courtesy of Mitutoyo America.

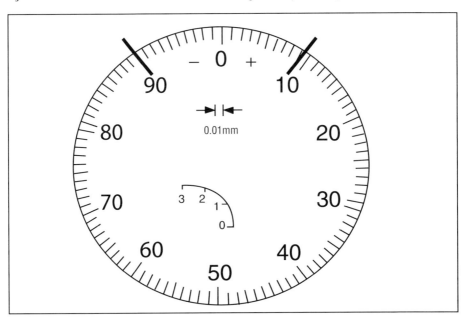

Figure 18-10: The part tolerance is best explained as ±10 increments instead of "from the 90 to the 10". Image courtesy of Mitutoyo America.

168   Exposing the Myths of Industrial Precision Measurement Control

± 0.1 specification, they could not get past the fact that the indicator did not have a 90 on the face. They actually could not be retaught to use the balanced indicator. I'm not joking, it really happens.

When I explain this situation to people in my instrument reading classes, I try and draw a parallel to learning to tell time using a clock. My wristwatch does not have numbers on the dial face (only slashes), and if you gave me a bogus watch as a gag gift that had the wrong numbers on the watch face I could still get where I needed to be on time. It's not the numbers that tell the time… it's the *resolution* and the *increments*.

Even this one goes back to two-steps.

Chapter 19
Putting it All Together

"Who controls the past controls the future…who controls the present also controls the past…" —George Orwell – Author of *"1984"*

Mr. Orwell was not really that far off in his epic novel about the year I graduated high school. He spoke of an age when the Government (whom he termed "Big Brother") would have surveillance equipment in our neighborhoods to keep society inline. He put his own spin on it, and when you think of how our credit and medical records are kept online, or how we can't buy a pack of chewing gum without a barcode scanner, it makes you realize how close he was in his vision of the future.

Inspiration can be defined as looking at the exact same thing that others look at, but seeing something different. I want to tell you a story that hits close to home for me. Several years ago, in my hometown of Portland, IN, one of the strongest industrial facilities was Teledyne Portland Forge. At that time a man named Greg Moser held the title of Vice President of Engineering, and was the obvious successor to become President of the facility when President Robert "Doc Bob" Read retired.

Greg had established a side business in 1982 in his garage. This business, Moser Engineering, started with shortening and reslpining the rear end of Greg's 1957 Oldsmobile racecar. After discovering that others were in need of this service, Greg began advertising his new business and perfecting the respining process. Eventually, this side business became a crossroads in Greg's business career. When he was finally offered the position of President of Teledyne Portland Forge, the directors felt Moser Engineering was a conflict of interest and added the stipulation that, to be President, he had to sell his engineering business.

Greg had a vision of what Moser Engineering could become with the right focus, planning, and execution. He also knew that people purchased products for three reasons: Quality, Price, and Delivery. You can have only one of the three and sell products. If you have super delivery time, a high price, and poor quality, people will buy your products. You can have a cheap price, terrible quality, and terrible delivery, and people will buy your products. Or, you can have a super high price, terrible delivery, but impeccable quality and people will buy your products.

Now, what do you think would happen if you could master two out of the three? And what would you have if you could develop and implement a strategy that allowed your system to attain all three? You'd most likely have a result similar to Moser Engineering. Greg set out to correct mistakes that he'd continually seen in industry. Part of this concept involved learning not only from his mistakes, but also the mistakes of others. Another philosophy of his is truly amazing: Never raise the price of products to pay for wage increases. Greg based pay wage increases on productivity improvements. This has allowed his products to remain at the same cost they were in the early 1980s, while Moser Engineering has grown several hundred percent and his employees are among the highest paid industrial employees in Jay County Indiana.

Moser Engineering entered the new after-market axle business and went from 0% to 70% of the market-share using a very simple concept: he wanted to be the best. Those who want their systems to be the best don't ask, "can't we just… and get by," they ask, "why don't we just… and lead the pack!" Greg got up early, studied more, and worked harder. Greg had a vision—he looked at the same thing others were looking at and saw something more. I'd bet my last dollar that many people looked at him in the early- to mid-1980s and thought this venture was **crazy**. I'd go double or nothing (so you can try and win your money back) that those same people would now use the term **brilliant**.

Figure 19-1: The original Moser Engineering facility. Images courtesy of Rob Moser.

Figure 19-2: The Moser Engineering facility today. Images courtesy of Rob Moser.

The word courage comes to mind, but with a twist of irony. Most would define courage as the act of conquering something that we are afraid of, so courage and fear seem to go hand-in-hand. If you've ever asked someone who you feel has performed what you consider a courageous act, they'll often tell you they feared living with themselves if they had not acted, much more than they feared taking the courageous action. So of the choices (acting or failing to act), the person chose the option they feared less. Maybe courage has more to do with the choices our conscience allows us to deal with than it does how much we fear the choice itself? I would bet that Greg Moser feared accepting the position of President of Teledyne Portland Forge (knowing that he would have to close his eyes and walk away from his vision) more than he feared the sacrifices that would be needed to turn his vision into reality. It doesn't get any more courageous than that.

As with Moser Engineering, the future of industrial measurement has yet to be determined. More than likely, what drives the future of industrial technology has already been invented but not yet viewed from the right perspective to get the full vision of what it will be able to do or where it can take us. One thing is certain, the ancient Aztecs and Egyptians built some of the most technologically difficult structures in the history of mankind by drawing images in the sand, and by discovering the value of Pi by rolling a jar across the ground. Even the computers we use everyday to calculate the algorithms used to measure roundness of surface finish in tenths- or hundredths- of a micron (approximately four-millionths of an inch) are calculating nothing more than a series of zeros and ones. If we really make the effort, it can be broken down to this level of simplicity.

As you lay the cornerstone in place at your industrial facility to build the foundation of your quality system in the form of a word-class inspection, measurement, and test equipment control system, just remember that you are performing a task that will sometimes have a safety net that resembles a concrete parking lot. When you inspect a fixture or calibrate an instrument that you determine to be within acceptable limits and fit for use on your shop's floor, there may be no one under your roof qualified to "have your back" if you make a mistake. When you give the instrument your seal of approval, it could be used fifty-times a day to measure hundreds of parts each week for three- to six-months before being inspected again. In some cases, there is no margin for error. It's a responsibility that you either accept or you don't. If you do, welcome to the club: you are a member of a very elite band of brothers and sisters.

In a production environment, 95% efficiency is considered excellent. When dealing with measurement instrument control and calibration, 99% (or one-mistake in two-weeks) may not be a passing grade. At the end of the day it all comes down to expectations: If you make sure that nobody will ever expect more out of you than you expect out of yourself, you are already half-way to a thorough measurement control system.

Now, back to Earth, never forget that with this rare task that you are undertaking, there may be few people that really understand some of the concepts. I've made the mistake of thinking that perhaps I didn't explain it right, or they would have bought into it. I later realized there would be a season and time for everything, and all one can do is be as prepared as possible. Prepare for battle and then pray for a peaceful resolution. Don't take this out of the chess context in which it is meant. Communication and understanding will always be the weakest link to advances, and there are always at least four-levels of understanding.

- ***Unconscious ignorance*** –

They do not know what they do not know.

- ***Conscious ignorance*** –

They know what they do not know, but don't know how to learn it.

- **Conscious understanding** –

They know what they know, but have to step their way through it.

- **Unconscious understanding** –

They know what they know on a different plane.

Hopefully, within two- to five-years along the learning curve you will be at a level of unconscious understanding within your measurement control system and the international standards that govern it. Unfortunately, you may likely be the only individual at your facility that has this understanding.

Some of the concepts discussed in this book—such as gage-to-gage variation, uncertainty, and others—represent phenomena that go against what has been established as the common wisdom of what is real and what is science fiction; what is present and what is absent within a measurement or measurement setup.

People at your facility will accept these concepts within their own time-frames. All you can do is be prepared to fill in the blanks when they're on the verge of becoming believers. These concepts are real and will be encountered (to some degree) at your facility. If your facility were a roulette wheel, these concepts would represent "the house." A standard roulette wheel has 36-colored (18-black and 18-red) spaces along with white spaces labeled 0 and 00. You could bet red all day long and I could stand beside you and bet black all day long and our odds would never be better than 18:38 each. In the long run the house always wins!

Quality in industry is never a sprint; it's a marathon, a very long run. If the opponents of these concepts have not experienced them to the degree needed to make them believers, they're just riding a lucky streak. Be patient and follow the fundamental laws and principles of these concepts. Eventually, the unbelievers will learn the only sure thing about luck: it always runs out!

Trust me, I've been there…

"We are usually convinced more easily by reasons we have found ourselves than by those which have occurred to others." Blaise Pascal, 1670.

Index

"If you can't convince them, confuse them!—Harry S. Truman

10:1 ratio (rule): 3, 32
5Ps: 47

Accreditation: 37, 38, 39, 141
Accuracy: 2-8, 18-21, 28, 31, 32, 35, 38, 40-43, 88, 100, 106, 109, 114-117, 123, 124, 135-139, 143, 146-148, 159
Ampere, defined: 10
Archimedes principal: 13
Artifact standard: 7, 9, 11, 13, 14
Average and range study: 77, 83

BIPM (Bureau International des poise y measures): 10
Bore gage: 143

Calibration: 1, 2, 4, 6-9, 11, 12, 14, 17-20, 22-25, 27, 29-34, 37-40, 42, 43, 47-49, 51, 52, 53, 60, 64-68, 81-83, 95, 96, 98, 100, 101, 109, 114-116, 119, 121, 125-132, 135-141, 146, 147, 150
Calibration intervals: 24, 33
Calibration label: 6, 30
Caliper: 3-5, 11, 18-20, 27, 28, 39, 55, 56, 81, 88, 104, 108, 162, 163
Candela, defined: 10
CMM monitoring: 66-68
Coefficient of thermal expansion: 55
Controlled procedures: 17

Document and data control: 18

Environmental variation: 90, 104
Equipment variation (EV): 73
Error: 4, 5, 7, 24, 32, 41, 75, 77, 79, 81, 82, 88, 95, 95, 97, 98, 100, 101, 108, 117, 128, 132, 133, 138, 143, 145, 147, 148, 163

Gage block: 9, 19, 20, 22, 27, 37, 39, 40, 43, 46, 54-57, 81, 88, 104-110, 136-138, 147, 148, 150-153
Gage block tolerances: 46
Gage Maker's Tolerance Chart: 40, 41
Gage R&R: *See Measurement System Analysis*
Grade (calibration rating): 42
Granite surface plates: 38, 44, 45, 56

Histogram: 63, 64, 77
Hydrometer: 13, 14

Indicator: 20, 27, 31, 32, 38-41, 47, 57, 76, 91, 92 100, 140, 142-144, 155, 156, 164-169
Instrument master list: 27
Instrument variation: 88
ISO-17025: 17, 37-39, 96, 107, 108
Inventory system (for measuring devices): 113

Kelvin, defined: 10
Kilogram, defined: 10

Linearity study: 77, 81

Margin of error (safe): 97
Mean (average): 65
Measurement System Analysis (Gage R&R): 21, 73, 76, 87
Meter, defined: 10
Micrometer: 2, 4, 5, 14, 15, 27, 28, 31, 39, 56, 57, 81, 82, 88, 93, 136-138, 163, 164
Mole, defined: 10

N.I.S.T.: 2, 9, 33, 38, 73-86, 90-93, 105, 106, 111, 123, 128
Normal distribution: 63, 101

Operator variation: 88

Part geometry: 89
Part variation: 92
Phonetic alphabet: 51
Pin gages: 22, 29
Precision: 3, 28, 32, 38, 50, 53, 74, 75, 77, 87, 93, 99, 108, 123, 132, 136, 138, 141, 149, 150, 159, 162
Probability: 64
Protocol standard: 12, 17, 25, 138
Purchase orders: 47

QS 9000: 1, 2, 6, 7, 8, 17, 22, 32, 33, 43, 53
Quality assurance: 135
Quality control: 135

R chart: 69
Random sample: 78
Range study: 77
Rectangular distribution: 101
Reference standard: 14, 64
Repeatability: 40, 73, 77
Reproducibility: 74, 77
Resolution (instrument): 31
Ring gages: 41

Second, defined: 10
Sigma: 63
SPC: 63-72
Specific gravity: 13
Stability study: 77
Standard deviation: 63

Thermal conductivity: 58
Thermal effects: 53-62, 90, 102, 104, 111
Thermal expansion (calculating): 54
Thermometer calibration: 11
Third party inspection: 17
Tolerance: 3
Traceability: 2, 9, 11
Training: 99, 133, 149-169
Triangular distribution: 101
TS-16949: 17, 73, 96, 113

Uncertainty: 1, 12, 37-39, 77, 88, 95-102, 103-111
Uncertainty contributors: 100, 103
Unique identification (gage) number: 30
U-shaped distribution: 101

Variation: 15, 23, 24, 31, 40, 63, 64, 67, 69-71, 73-90, 92, 93, 97-101, 105-107, 110, 111, 133, 139, 141, 142, 144, 145, 147, 148, 151, 152

Work instructions: 17, 18, 99, 139

X-bar chart: 69